S0-CAH-364

ALLE ZEIT WACH
1842

E. Ottow · K. Schöllkopf · B.-G. Schulz (Eds.)

Stereoselective Synthesis

Lectures honouring

Prof. Dr. Dr. h.c. Rudolf Wiechert

With 179 figures, 38 of them in colour

Springer-Verlag
Berlin Heidelberg New York
London Paris Tokyo
Hong Kong Barcelona Budapest

Dr. Eckhard Ottow
Dr. Klaus Schöllkopf
Dr. Bernd-Günter Schulz
Schering AG Berlin
D-13342 Berlin

ISBN 3-540-57202-3 Springer Verlag Berlin Heidelberg New York
ISBN 0-387-57202-3 Springer-Verlag New York Berlin Heidelberg

Library of Congress Cataloging-in-Publication Data
Stereoselektive synthesis: lectures honouring Prof. Dr. Dr. h.c. Rudolf Wiechert/E. Ottow, K. Schöllkopf, B.-G. Schulz (Hrsg.). p.cm. Papers from a one-day symposium in honor of Wiechert's 65th birthday. Includes bibliographical references and index.
ISBN 3-540-57202-3 (Berlin: acid-free).
ISBN 0-387-57202-3 (New York: acid-free)
1. Stereochemistry-Congresses. 2. Organic compounds-Synthesis-Congresses. I. Wiechert, Rudolf, 1928-. Ottow, E. (Eckhard), 1953-. III. Schöllkopf, K. (Klaus), 1953-. IV. Schulz, B.-G. (Bernd-Günter), 1950-.

Printed in Germany

Dataconversion: Lewis & Leins, Berlin; Printing: Saladruck, Berlin;
Binding: Lüderitz & Bauer, Berlin
51/3020-5 4 3 2 1 - Printed on acid-free paper

Preface

Nature is a never ending source of interesting chiral molecular targets supplying the world of synthetic organic chemists with puzzling structural elements together with demanding stereochemical features. In addition, due to their biological activity various of these molecules have proved of great interest to mankind. Traditionally, natural products have been and are used themselves as drugs or as lead compounds in pharmaceutical research. In Medicinal Chemistry, struggling for rational ways from leads to more potent and selective active ingredients, the responsibility for creating stereochemically well defined and enantiomerically pure molecules becomes more and more important. Different, or even opposite, biological activity of two enantiomers, potentially leading to undesired adverse effects or reduction of efficacy has become a major issue in drug development and approval. Although nature handles the mystery of chirality with ease, using a catalytic asymmetric machinery called enzymes, organic chemists have still a long way to go to reach the same efficiency and diversity. Therefore, for preparative purposes the need and desire for new and highly stereoselective reactions is steadily increasing. As an ultimate goal the design of man-made chiral catalysts which mimic natural enzymes in their efficiency in converting prochiral centers to chiral ones, but which offer optional chirality is evident.

With this in mind, and in view of his significant contributions to Medicinal Chemistry, Schering AG has taken the opportunity to honour the scientific and professional work as well as the personality of Prof. Dr. Dr. h.c. Rudolf Wiechert, head of central research in chemistry and molecular biology. On the occasion of his 65th birthday a one day symposium on stereoselective synthesis was organized. Six distinguished speakers were invited to give a broader overview about recent developments in the field. In addition to the excellent lectures presented the elegant chairing by Prof. Dr. Helmut Schwarz of the Technische Universität Berlin made a major contribution to the success of this symposium.

Finally, thanks are due to all who helped with the technical organization of the symposium, particularly Ms. Hans, Ms. Rausch, Ms. Schulz and Ms. Zahn. In addition, the help for proof-reading by Dr. A. Cleve is gratefully acknowledged

Berlin, in September 1993

E. Ottow
K. Schöllkopf
B.-G. Schulz

Contents

CVs of the Lecturers

EKKEHARD WINTERFELDT was born in Danzig in 1932. He received his degrees from the Technische Universität Braunschweig, where he finished his dissertation under the direction of Prof. F. Bohlmann in 1958. He moved to the Technische Universität Berlin, obtained his habilitation in 1962 and in 1967 he was promoted to associate professor. Since 1970 he has held his current position as Professor of Organic Chemistry at the Universität Hannover. His research is devoted to natural product synthesis with emphasis on biomimetic approaches and stereoselective transformations.

ANDREAS PFALTZ was born in Basel, Switzerland, in 1948. He studied at the ETH in Zürich where he completed his Ph.D. thesis in 1978 under the guidance of Prof. A. Eschenmoser. After a postdoctoral stay at Columbia University with Prof. G. Stork, he returned to the ETH. There he began his independent research and was appointed private lecturer in 1987. In 1990 he joined the faculty at the Universität Basel as Professor of Organic Chemistry. His research interests are in the areas of synthetic organometallic and bioorganometallic chemistry, with special emphasis on asymmetric catalysis.

JOHANN MULZER was born in Prien, Germany in 1944. He received his doctoral degree from the Universität München in 1974 under the direction of Prof. R. Huisgen. He spent one year as postdoctoral fellow at Harvard University with Prof. E. J. Corey working on the LHASA-project. He went back to the Universität München where he obtained his habilitation in 1980. In 1982 he was appointed professor at the Universität Düsseldorf. Since 1984 he has held his present position as Professor of Organic Chemistry at the Freie Universität Berlin. His research interests are in the field of stereoselective synthesis of natural products and the development of new synthetic methods.

DIETER ENDERS was born in Butzbach, Germany, in 1946. He studied chemistry at the Universität Gießen, where he received his doctoral degree in 1974 under the guidance of Prof. D. Seebach. He spent one postdoctoral year at Harvard University with Prof. E. J. Corey and returned to the Universität Gießen, where he received his habilitation in 1979. In 1980 he was appointed professor at the Universität Bonn and since 1985 he has held his current position as Professor of Organic Chemistry at the Technische Hochschule Aachen. His research interests include asymmetric synthesis, new synthetic methods using organometallics and the synthesis of biologically active compounds.

REINHARD W. HOFFMANN was born in 1933 in Würzburg. He studied chemistry at the Universität Bonn and received his Dr. rer. nat. in 1958 under the direction of Prof. B. Helferich. He spent two years as postdoctoral fellow at The Pennsylvania State University with Prof. G. W. Brinkley and moved then to the Universität Heidelberg for another postdoctoral year with Prof. G. Wittig. In 1964 he obtained his habilitation and in 1967 he moved as lecturer to the Technische Hochschule Darmstadt and in 1970 to his present position as Professor of Organic Chemistry at the Universität Marburg. His research interests are new methods in stereoselective C-C-bond formation, their application in natural products synthesis and the stereochemistry of reactive organometallic compounds.

GERHARD QUINKERT was born in Lüdenscheid in 1927 and obtained his scientific education at the Technische Hochschule Braunschweig. He received the Dr. rer. nat. in 1955 under the guidance of Prof. H. H. Inhoffen. Subsequently, he spent two years as postdoctoral fellow with Prof. H. H. Inhoffen and Prof. D. H. R. Barton at Imperial College London, respectively. Then he came back to Braunschweig and became lecturer in 1961 and professor in 1963. In 1970 he moved to his present position as Professor of Organic Chemistry at the Universität Frankfurt am Main. His main research interests are in the field of organic photochemistry and the stereoselective total synthesis of bioactive compounds.

Abbreviations

AIDS	Acquired Immunodeficiency Syndrome
AZT	3'-Azido-3'-deoxythymidine
BAZ	Benzodiazepinoalanine = [*R*-(*R**,*S**)]-α-amino-2,3-dihydro-2-oxo-5-phenyl-1*H*-1,4-benzodiazepine-3-propanoic acid
L-BMAA	L-β-(Methylamino)alanine
BOC	*Tert*-butoxycarbonyl = (1,1-dimethylethoxy)carbonyl
BOM	Benzyloxymethyl = (phenylmethoxy)methyl
BSA	*N*,*O*-Bis(trimethylsilyl)acetamide = *N*-(trimethylsilyl)ethanimidic acid trimethylsilyl ester
BuLi	Butyllithium
DAHP	3-Deoxy-D-*arabino*-2-heptulosonic acid 7-phosphate
DDC	2',3'-Dideoxycytosine
DDI	2',3'-Dideoxyinosine
DDQ	2,3-Dichloro-5,6-dicyano-2,5-cyclohexadiene-1,4-dione
DHAP	Dihydroxyacetone phosphate
DIOP	2,3-*O*-isopropylidene-2,3-dihydroxy-1,4-bis(diphenylphosphino)butane = [(2,2-dimethyl-1,3-dioxolane-4,5diyl)bis (methylene)]bis[diphenylphosphine]
FDP-aldolase	Fructose 1,6-diposphate aldolase
Fmoc	9-Fluorenylmethoxycarbonyl
Fmoc-OSu	9-Fluorenylmethoxycarbonyl-*N*-hydroxysuccinimide = 1-[[(9-fluorenylmethoxy)carbonyl]oxy]-2,5-pyrrolidinedione
FSH	Follicle stimulating hormone
GBF	Gesellschaft für Biotechnologische Forschung
HIV	Human immunodeficiency virus
HMP	3-Hydroxy-4-methyl-proline
KDG	2-Keto-3-deoxygluconate = 3-deoxy-D-2-gluculosonic acid
KDO	3-Deoxy-D-*manno*-2-octulosonic acid = 2-keto-3-deoxyoctonate

LDA	Lithium diisopropylamide
LH	Luteinizing Hormone
LHRH	Luteinizing Hormone Releasing Hormone
MAD	Methylaluminumbis[2,6-di(*tert*-butyl)-4-methylphenoxide]
MOM	Methoxymethyl
MorLys	(2*S*)-2-Amino-6-(4-morpholinyl)hexanoic acid
(S)-MTPA	(*S*)-α-Methoxy-α-(trifluoromethyl)benzeneacetic acid
NOE	Nuclear Overhauser effect
PEP	Phosphoenolpyruvic acid
PipLys	(2*S*)-2-Amino-6-(1-piperidinyl)hexenoic acid
Phth	Phthalimido
PyrLys	(2*S*)-2-Amino-6-(1-pyrrolidinyl)hexanoic acid
RAMP	(*R*)-1-Amino-2-(methoxymethyl)pyrrolidine
SAEP	(*S*)-1-Amino-2-(1-ethyl-1-methoxypropyl)pyrrolidine
SAMP	(*S*)-1-Amino-2-(methoxymethyl)pyrrolidine
TADDOLs	α,α,α',α'-Tetraaryl-1,3-dioxolan-4,5-dimethanols
TEMPO	2,2,6,6-tetramethylpiperidinooxy, free radical
TFAA	Trifluoroacetic anhydride
TMS-SAMP	(*S*)-1-[(Trimethylsilyl)amino]-2-(methoxymethyl)pyrrolidine

Steroids and Stereochemistry

E. WINTERFELDT

Institut für Organische Chemie der Universität Hannover,
Schneiderberg 1 B, D-30167 Hannover

Abstract. The key role of steroids in the development of stereochemistry is demonstrated with a number of stereospecific and stereoselective transformations. The crucial role of conformations is highlighted with examples from the fields of cycloaddition and complex formation.

Owing to their well defined configuration and in many cases rigidly fixed conformation, steroids proved to be the substrates of choice at all important stages of development and progress in stereochemistry. Whenever any important rules or principles in either static or dynamic stereochemistry were revealed final convincing results were generally obtained from well planned experiments with steroids and whenever new stereoselective or stereospecific reaction pathways were proposed steroids offered themselves as the ideal testing ground. Satisfying and convincing results here very often paved the road to successful applications everywhere. To start with the very spectacular and far reaching contributions to

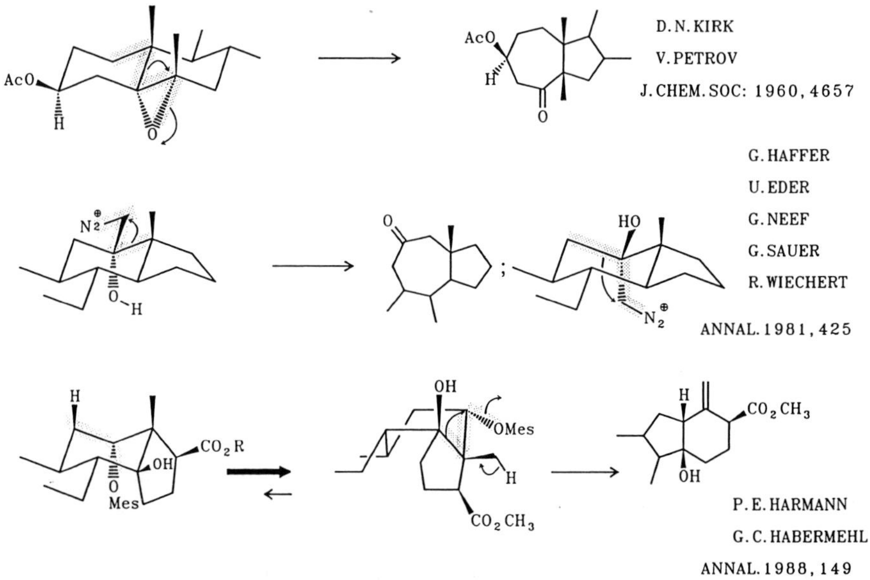

D. N. KIRK
V. PETROV
J. CHEM. SOC: 1960, 4657

G. HAFFER
U. EDER
G. NEEF
G. SAUER
R. WIECHERT
ANNAL. 1981, 425

P. E. HARMANN
G. C. HABERMEHL
ANNAL. 1988, 149

Scheme 1

Stereoselective Synthesis
Editors: Ottow, Schöllkopf, Schulz

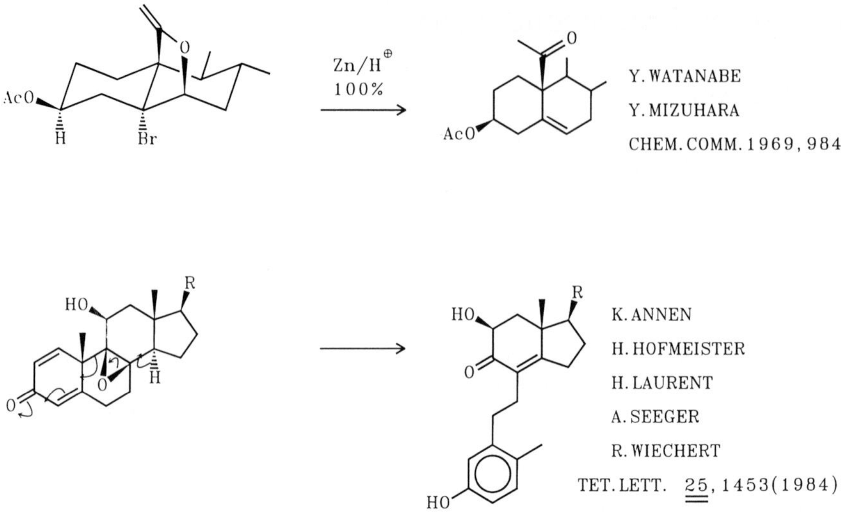

Zn/$H^{\oplus}$
100%

Y. WATANABE
Y. MIZUHARA
CHEM. COMM. 1969, 984

K. ANNEN
H. HOFMEISTER
H. LAURENT
A. SEEGER
R. WIECHERT
TET. LETT. 25, 1453 (1984)

Scheme 2

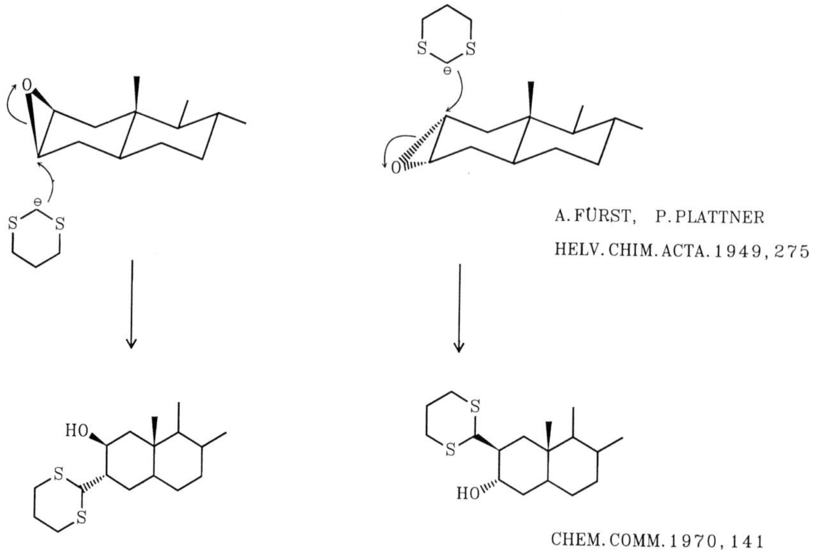

A. FÜRST, P. PLATTNER
HELV. CHIM. ACTA. 1949, 275

CHEM. COMM. 1970, 141

Scheme 3

the area of conformation analysis one has of course to mention the pioneering work of *Sir Derek Barton* who guided organic chemists into the realm of non bonded interactions that determine conformations and made them realize the important differences between axial and equatorial substituents as far as steric hindrance and reaction rates are concerned. The rules and generalizations that emerged from these efforts did not only convincingly explain numerous earlier

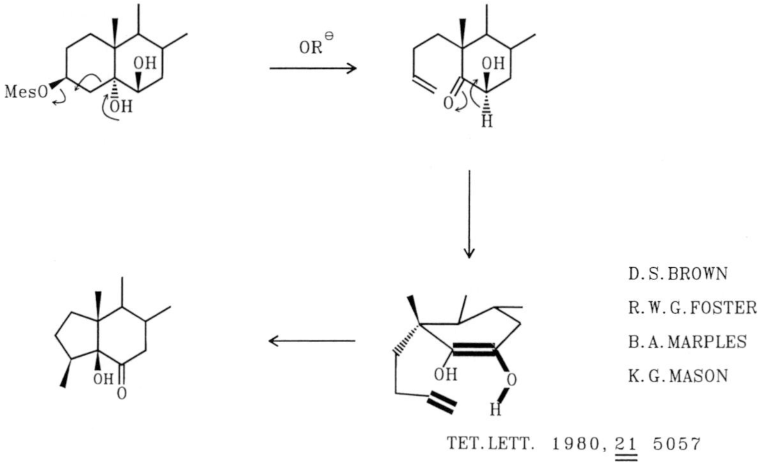

Scheme 4a

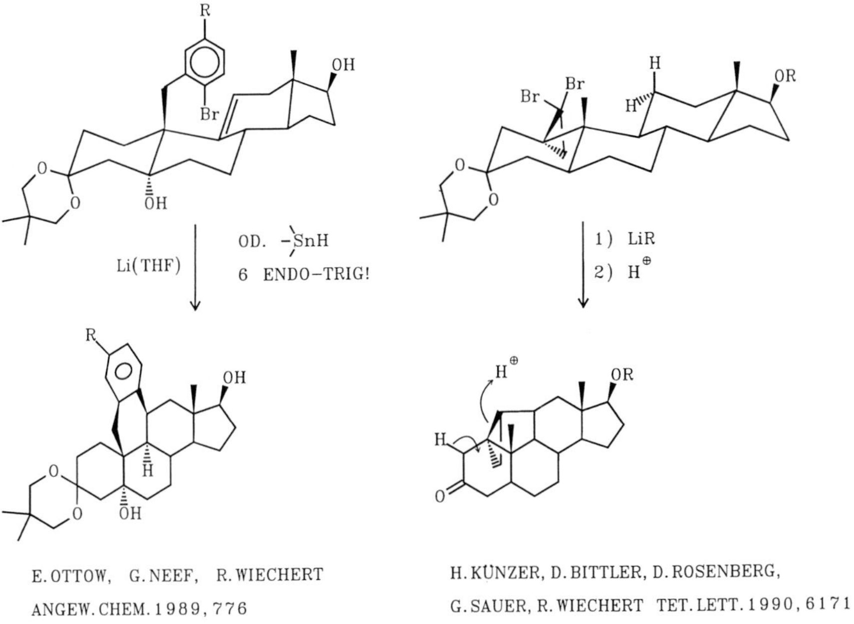

Scheme 4b

purely empirical observations in alicyclic chemistry in general, but did also lay a very firm foundation for the development of the stereoelectronic principle which calls for optimal bond overlap for transformations in rigid systems. Besides rearrangements (**Scheme 1**) and fragmentations (**Scheme 2**) the highly selective ring opening of three membered rings is a very typical result in this area with far reaching consequences for the mechanistic interpretation as well as synthetic use of regioselective manipulations of these cyclic systems. As far as epoxides are concerned this has been cast into the *Fürst Plattner*-rule, which allows for very safe predictions concerning the regioselectivity of the strictly nucleophilic attack to this moiety (**Scheme 3**). As these processes rely on the rigidity and conformational reliability of the ring system one has in this context also to mention numerous neighbouring group effects that have been noticed in the steroid field, giving in some cases rise to very special or unexpected products like the ones portrayed in **Scheme 4a/Scheme 4b** and which culminate in through space oxidation of non-activated methyl groups like the *Barton*-photolysis of nitrites or in the remote oxidation techniques investigated by *Breslow* and his colleagues which led to an interesting application in the Schering company (**Scheme 5**).

While the angular methyl groups of steroids represent a very typical case of stereoselection owing to inert volume or passive space demand steroids do of course also offer cases of highly efficient directing by active volume as for instance in the perfect stereoselection in *Simmons-Smith* cyclopropanations triggered by complex formation to a properly placed hydroxy group or as in a recent case of ortho-ester formation (**Scheme 6a/Scheme 6b**) while the configurational control as exercised by the two angular methyl groups is of course considered to be responsible for the prevailing α-attack to steroids there are well established examples for conformational control, too. In the ring A/B cis series for instance highly selective β-attack is noticed in carbonyl reductions (**Scheme 7**) and judging from the perfect concave-convex situation in the ring A/B-area one is not at all surprised to learn that this type of stereodirection generally overrules any configurational control. Without going into detail one should realize at this point

R. BRESLOW
R. CORCORAN
J. A. DALE
S. LIU
P. KALICKY
J. AMER. CHEM. SOC. 1974, 1973

U. KERB
M. STAHNKE
P. E. SCHULZE
R. WIECHERT
ANGEW. CHEM. 93 89(1981)

Scheme 5

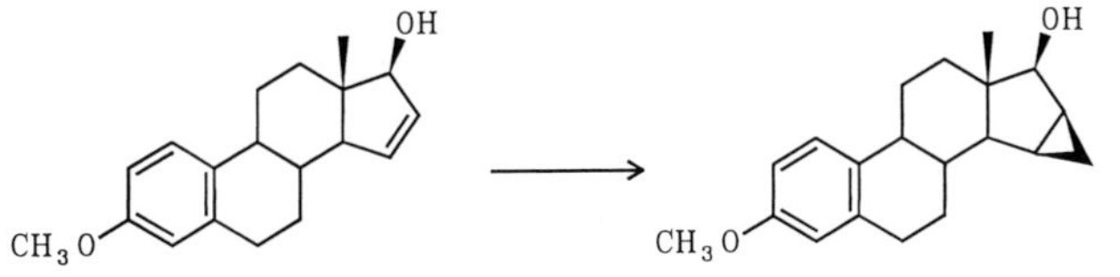

CH$_3$O — OH → CH$_3$O — OH

G. SCHMIDT
K. PREZEWOWSKI
G. SCHULZ
R. WIECHERT
CHEM. BER. 101, 939 (1968)
S. A. CHEM. BER. 106, 888 (1973)

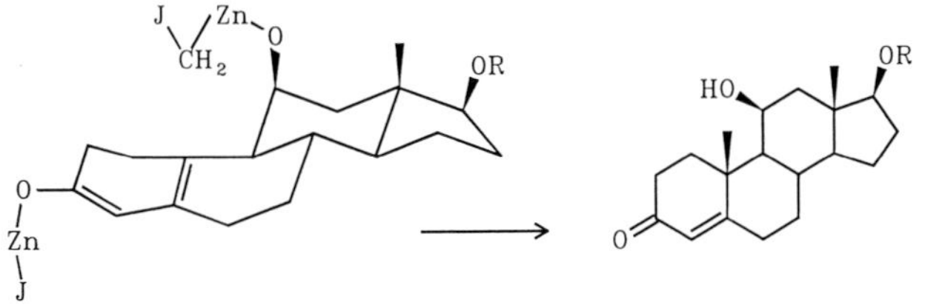

J–Zn–O, CH$_2$, OR, O–Zn–J → HO, OR, O

P. TURNBULL
K. SYHORA
J. H. FRIED
JACS 88, 4764 (1966)

Scheme 6a

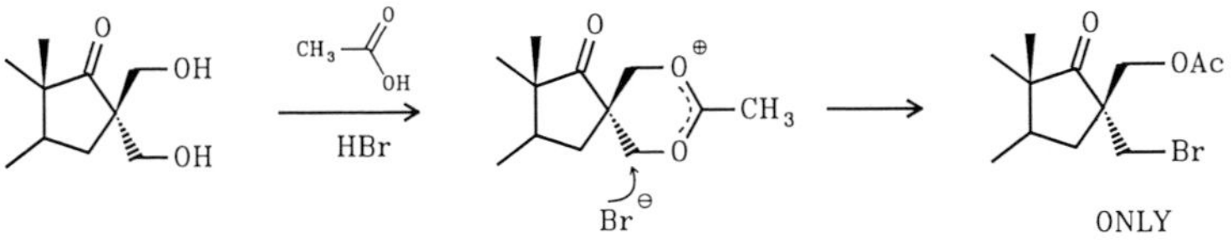

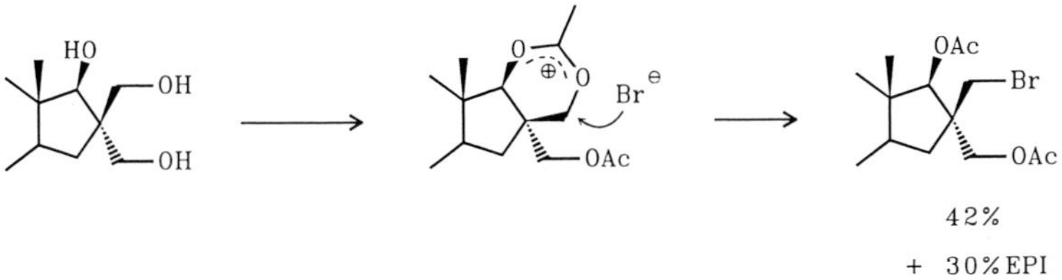

O, OH, OH; CH$_3$COOH, HBr → O, O$^{\oplus}$, CH$_3$, O, Br$^{\ominus}$ → O, OAc, Br

ONLY

HO, OH, OH → O, $\oplus$, O, Br$^{\ominus}$, OAc → OAc, Br, OAc

42%
+ 30% EPI

P. SOHAR, A. FÜRJES, J. WÖLFLING, G. SCHNEIDER
SYNTHESIS 1992, 1280

Scheme 6b

that even subtle conformational changes may be reflected in quite distinct differences in reaction behaviour and to me there is no doubt that with growing knowledge about highly populated low energy conformations quite a number of unexpected or unusual transformations may find quite satisfying explanations along these lines.

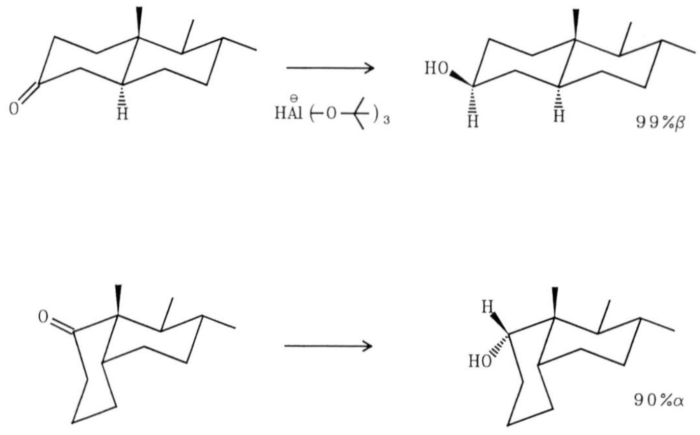

Scheme 7

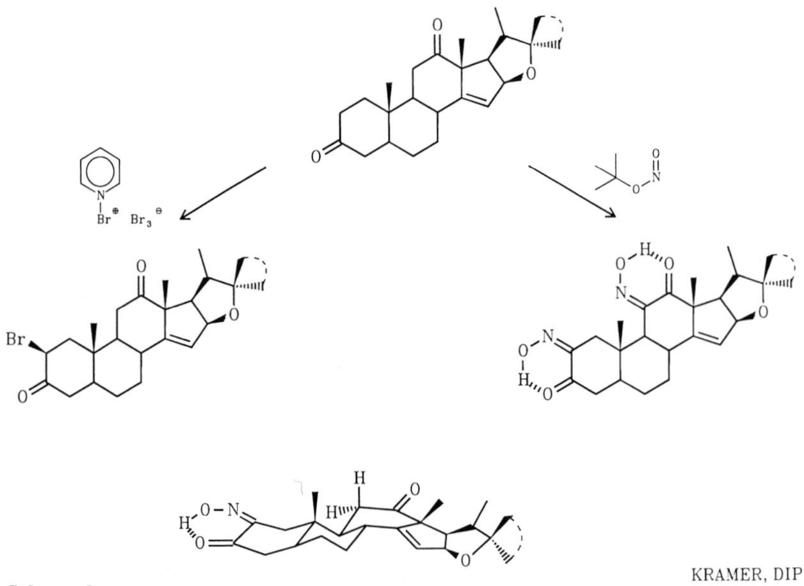

KRAMER, DIPL. ARB.

Scheme 8

Although there are certainly many, suffice it to cite just one example from our own work (**Scheme 8**) indicating the conformational changes caused by Δ14,15 unsaturation. While the corresponding saturated material on base or acid catalysed nitrosation with *tert*-butylnitrite gives exclusively rise to the C_2-monooximes expected, the Δ14,15 diketone suffered immediate bisnitrosation to form a bisoxime in high yield and a rough comparison pointed at very similar rates for both reactions. MMX and PCmodel investigations indicated a substantial angle change for the two normally nearly parallel bonds of the C_{18} (10°) and C_{19} (6°) methyl groups, thus widening the gap between these two to an extent which allows for electrophilic attack at C_{11}.

OAc
CH$_3$O
EWG
J.R.BULL
R.J.THOMSON
H.LAURENT
H.SCHROEDER
R.WIECHERT
GER.OFF.3628189P
1
OAc
EWG
CH$_3$O
OAc
EWG
2

Scheme 9

Certainly a quite spectacular effect of conformational control is noticed with the excellent β-selectivity of *Diels-Alder* cycloadditions to ring D cyclopentadienes, as initially described by *Solo*, further investigated by *Bull* and imaginatively applied at the *Schering Research Laboratories* (**Scheme 9**). Adducts of type **2** were elaborated into various directions by either hydrogenation or oxidative fragmentation of the remaining double bond.

Even a cursory look at conformation **1** of the steroid C/D-segment explains the preferential β-approach immediately, as α-attack into the concave area can safely be excluded.

An inspection of this steroid substructure encouraged us to synthesize the enantiomerically pure dienes from the easy to make *Hajos-Wiechert* ketone (**Scheme 10**), which offered themselves as highly selective chiral templates which can be used in three different ways.

Firstly, it may combine with a symmetric prochiral dienophile like for instance quinone (**Scheme 11**) to form **4**, which then is transformed in a highly diastereoselective and (or) regioselective manner to form an adduct which in the following retro-*Diels-Alder* process splits into the diene and a chiral building block. Secondly, one can use the diene for chiral recognition with racemic dienophiles to achieve a kinetic resolution and this way one may even prepare both enantiomers of a chiral intermediate. In particularly lucky cases one may even trigger an additional thermal transformation in combination with the retro-process as indicated in **Scheme 12**. In a third application the diene may even serve for a combination of chiral recognition and molecular recognition, which in a cycloaddition reaction does amount to topological resolution as described in **Scheme 13**.

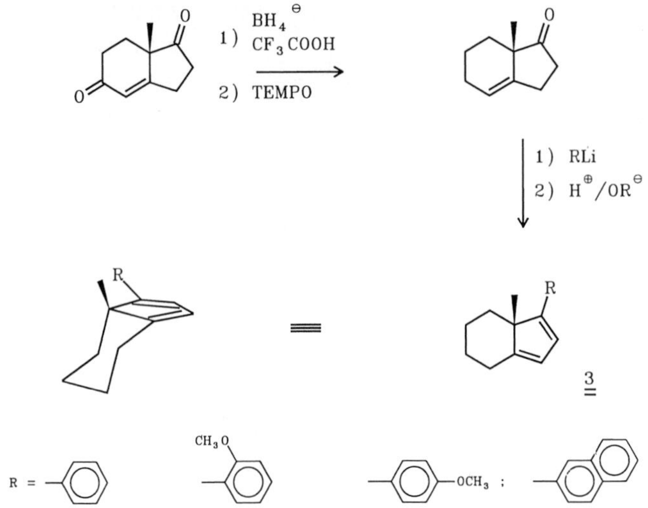

1) $BH_4^{\ominus}$ CF_3COOH

2) TEMPO

1) RLi

2) $H^{\oplus}/OR^{\ominus}$

R

3

R = phenyl; CH_3O-phenyl; phenyl-OCH_3; naphthyl

Scheme 10

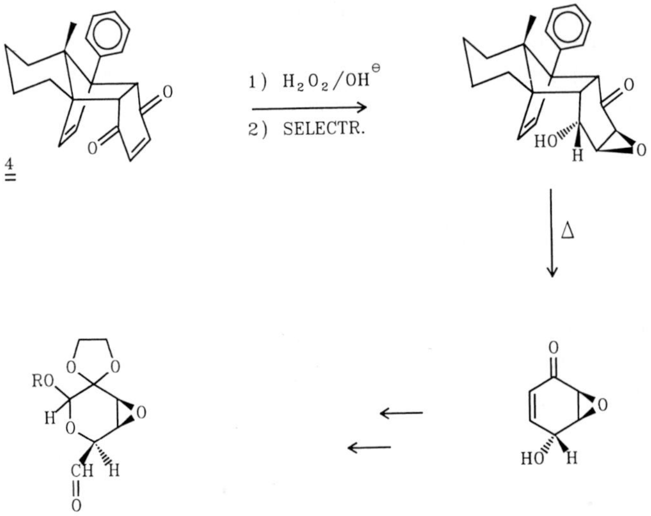

4

1) $H_2O_2/OH^{\ominus}$

2) SELECTR.

HO H O

Δ

HO H

RO H CH H O

Scheme 11

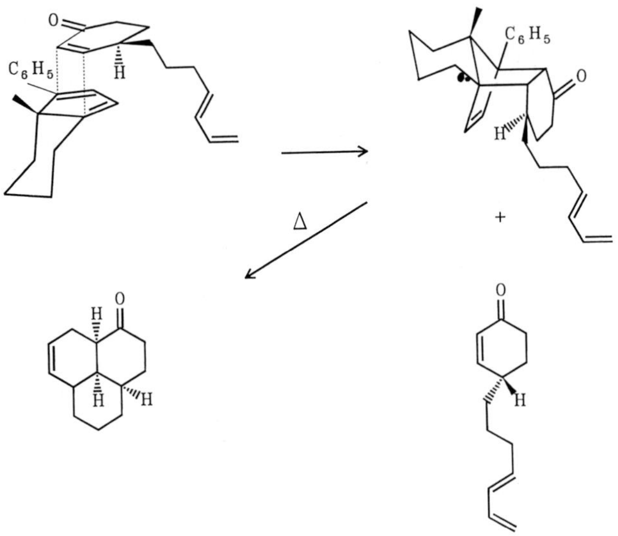

Scheme 12

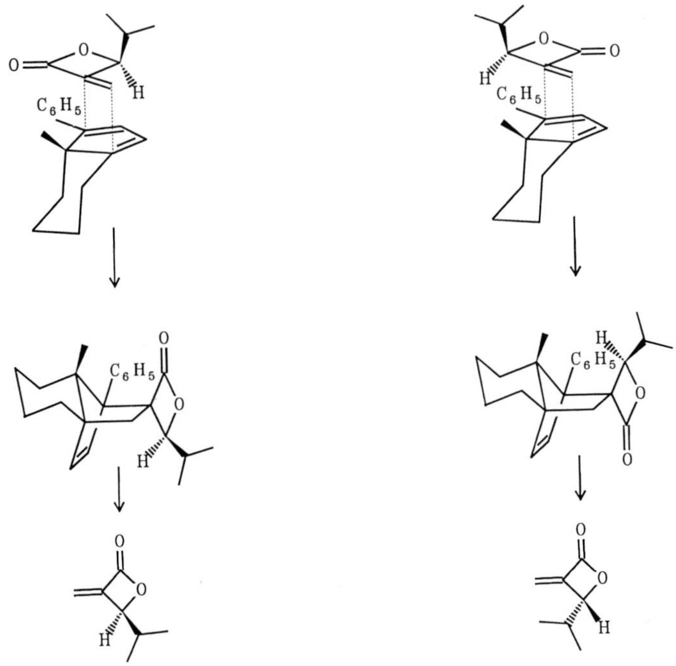

Scheme 13

With this cycloaddition we report the first example of topological resolution. In this case there is no knowledge of the reaction rates which may be very similar indeed if not even identical. Both enantiomers do in principle add to the diene but the one in the exo-fashion while the other one, also in order to keep the bulky isopropyl group away from the diene substituents, prefers the endo-route.

As these adduct isomers may easily be separated by chromatography one can again make use of the thermal retro-process to prepare pure enantiomers.

This way the stereochemical results obtained from steroids triggered a complete line of research that did eventually lead to the quite useful process of topological resolution.

In connection with these investigations a quite unusual observation was made, however, which called again for experimental advice from the steroid field as the final authority to settle a stereochemical question, which arose in the preparation of transition metal complexes. These transformations have attracted quite some attention and were applied to steroid chemistry for mainly two reasons. First of all diene complexes may serve as a diene protecting group, but additionally if they are formed via the usual α-attack they should direct subsequent transformations to the β-side thus changing the stereochemical preferences, as demonstrated in **Scheme 14**. When dienes of type **3** (**see Scheme 10**) were treated with iron carbonyls to form the usual complexes we obtained a mixture of diastereomeric reaction products (ratio: 1.4 : 1), which showed a very characteristic and quite revealing difference in the chemical shift for the resonance of the angular methyl group.

While the major, kinetically controlled reaction product showed the signal at 1.60 δ, the thermodynamically stable complex, which was obtained either at high reaction temperature or on subsequent heating of the low-temperature product had this very characteristic three proton singulet at 0.99 δ (starting material: 1.19). According to investigations by *von Philipsborn* and his colleagues with rigid well defined dienes, this last resonance proves the antiorientation of the cobalt moiety and the angular methyl group, while the downfield shift of the first one is caused by the synrelationship of these two groups.

These data call for configuration **5** for the kinetically controlled and for configuration **6** for the thermodynamically stable iron-carbonyl complex (see **Scheme 15**).

This assignment caused some concern at the beginning, as with 2π-4π cycloadditions one had never, neither on forcing reaction conditions nor on subsequent heating of adducts seen any convincing proof for α-attack of the dienophile. As a consequence we first of all retreated to further NMR-data, as one may safely predict that the downfield shift observed for the angular methyl group in **5** should be noticed in the same way - if not even stronger - with the axial protons H_A and H_B in configuration **6**, as these dip directly into the iron-carbonyl moiety. This was born out by NMR-experiments by *Dr. Michl* and *Miss Mayer* at the *Schering Company* and by *Dr. V. Wray* at the *GBF* in Stöckheim and which proved very clearly that these signals show up at 1.08-1.24 δ with the starting material **3** and are shifted downfield to H_A = 1.55 δ; H_B = 2.58 δ! with the endo-configuration **6**. These data leave us with the fact that

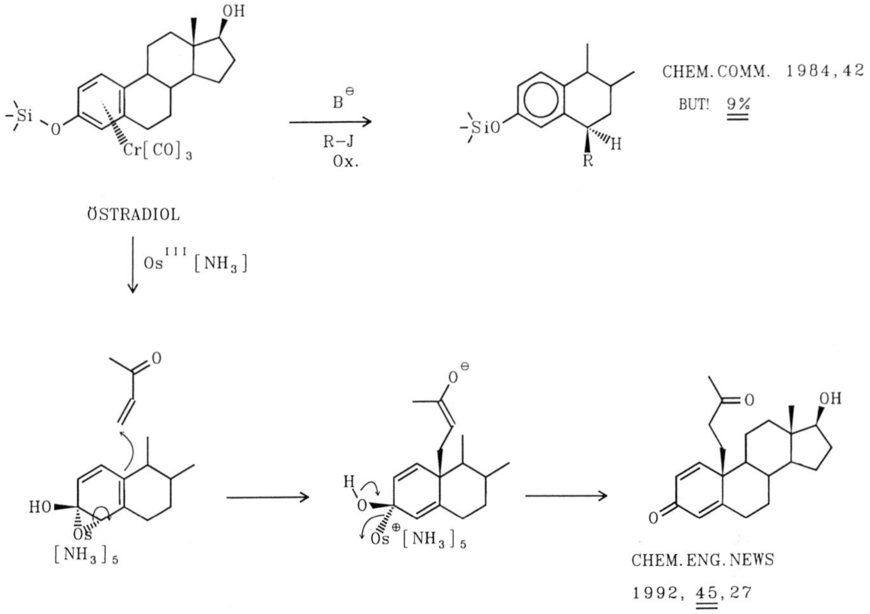
OH
Cr[CO]3
B⊖
R–J
Ox.
SiO
H
R
CHEM.COMM. 1984,42
BUT! 9%
ÖSTRADIOL
Os III [NH3]
O
HO
Os
[NH3]5
H
O
O⊖
Os⊕[NH3]5
O
OH
CHEM.ENG.NEWS
1992, 45,27

Scheme 14

C6H5
Fe[CO]3
HB
HA

5 6

C_6H_5	1,60	0,99	
P–OCH_3	1,61	0,97	
		H_A/H_B	1,55/2,58
		EDUCT	1,08–1,24

Scheme 15

obviously complex formation with iron-carbonyls differs very much from cycloadditions. Although kinetic control again leads to β-attack to form **5**, the thermodynamically stable species embeds the iron-carbonyl into the concave α-face of the molecule.

This is an atypical case where again one turns to steroids for help and advice with a stereochemical problem. As *Diels-Alder* additions as well as iron-carbonyl complex formation have been studied to quite some extent with ergosterol, we picked dienes of this type for comparison. We realised very quickly, however, that for many *Diels-Alder* additions described in the literature α-attack had simply been taken for granted and that additionally the *Diels-Alder* adducts had generally been formed under rather forcing conditions, which raised the question of thermodynamic versus kinetic control. To be sure in this respect we chose room-temperature, high-pressure cycloaddition of propargylic aldehyde to diene **7**, a reaction we were well acquainted with and which provides adduct **8** in nearly quantitative yield with excellent selectivity. NOE-experiments with this adduct provided firm evidence for exclusive α-addition of the dienophile, as convincing effects were noticed with H_A and H_B on irradiation at C_{19}, while, as one should expect, only the H_A-resonance responded on irradiation at C_{18}. These data prove that at least the kinetically controlled *Diels-Alder* reactions of activated triple bonds to the ring-B diene are highly α-face selective. Unfortunately, these compounds provide no information on the thermodynamically stable adducts, as even mild heating of these compounds results in a fast and irreversible retro-*Diels-Alder* process (see dotted line) to form aromatic ansa-compounds (**Scheme 16**).

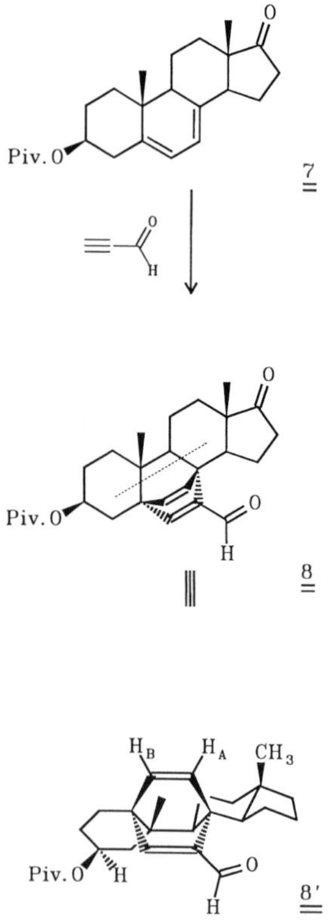

Piv.O 7 8 8' H_B H_A CH_3

Scheme 16

Being on safe ground with the question of face selectivity in cycloadditions we turned to the stereochemistry of complex formation.

This reaction has of course been investigated before but we realized quite soon that face selectivity and kinetic versus thermodynamic control had obviously not been major issues in these efforts. These reports generally focus on one complex to which the α-configuration is assigned. We therefore chose the mildest reaction conditions possible for this complexation (140°C, 5 h) and were pleased to note that ergosterol-acetate and $Fe[CO]_5$ under these conditions did form two iron-carbonyl complexes in a 90 : 10 ratio. Again NMR-data of these stereoisomers differed characteristically with respect to the C_{19}-resonance. While the major product gave rise to a 0.71 δ singulet, the minor one showed this signal at 1.19 δ (educt 0.64). When compared to the resonances noticed with complexes **5** and **6** and to the data reported by *von Philipsborn* one had to assign configuration **10** to the major and β-complex orientation **9** to the minor reaction product.

To address the question of thermodynamic stability the mixture described above was heated to 140°C for 30 hours. This lead to α-complex **10** exclusively thus pointing again at **9** as the result of a kinetically controlled β-addition to the 4π-system (**Scheme 17**).

These data from ergosterol and ergosterol-type compounds prove very clearly that approach control in iron-complex formation differs markedly from the one in *Diels-Alder* cycloadditions. While here all examples investigated so far call for an α-approach of the dienophile to form products like **8** with remarkable regioselectivity, complex formation does at least to some extent take place from the β-side of the molecule.

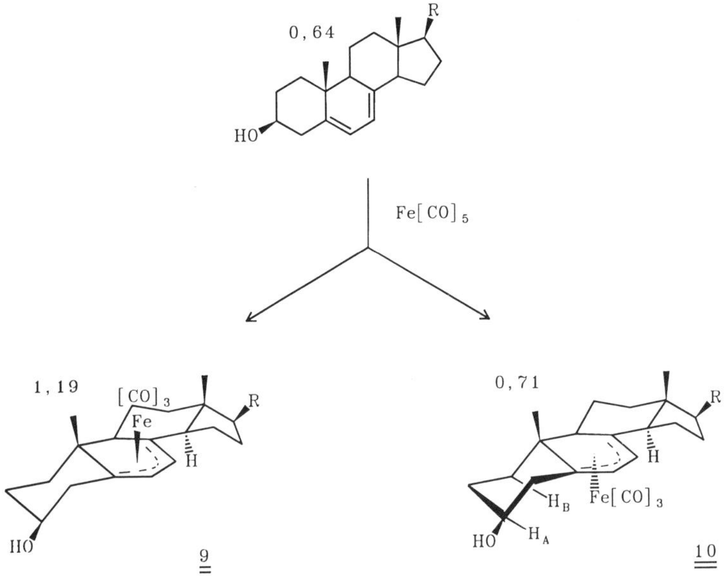

Scheme 17

Even if the thermodynamic stability of the α-complex in the steroid series is in line with results from cycloadditions, one is left with the quite remarkable observation that this very configuration also turns out to be the stable one with hydrindane complexes, although in this series *Diels-Alder* cycloadditions proved to be completely β-selective. This raises of course the question for the reasons of the higher stability of **10** and particularly of **6**. One item both these configurations have in common is the close vicinity of the axial C-H-bonds H_A and H_B to the iron-carbonyl moiety, and as iron complexes are known to interact with and even to activate C-H bonds, there may be a stabilizing effect from these interactions, too.

If this is true one has to draw the conclusion that groups, which are generally classified as inert volume and are normally just causing steric repulsion may in carbonyl-complex formation switch over to active volume and thus give rise to stabilizing interactions.

This will of course have to be thoroughly investigated and again there will probably be no better compounds than steroids, to be chosen as the testing ground for this phenomenon. At least in our laboratory the decision has been made to proceed along these lines. This proves additionally that those who some time ago predicted that steroid chemistry is approaching a dead end are very certainly wrong. It is still going strong with various aspects of pharmaceutical research, as may easily be judged from recent achievements in this company, but it will also for a long time to come continue to be the final authority in discussions and decisions on stereochemical problems. An old roman sentence referring to the central role of this empire in law and jurisdiction may be slightly changed to describe the situation: Steroids locuta - causa finita.

Enantioselective Catalysis with Chiral Metal Complexes

A. PFALTZ

Institut für Organische Chemie der Universität, St. Johanns-Ring 19,
CH-4056 Basel (Switzerland)

1. Introduction

Chemists have always been fascinated by the ability of enzymes to distinguish between two enantiomers or two enantiotopic sites in a molecule. For a long time, the almost perfect enantioselectivities observed in enzymatic reactions were considered beyond reach for non-biological catalysts. However, with the development of chiral rhodium-phosphine complexes for enantioselective hydrogenation [1] and the discovery of the proline-catalyzed Robinson annelation by Eder, Sauer, and Wiechert [2] and, independently, by Hajos and Parrish [3] around 1970, it became evident that high levels of enantioselectivity can also be achieved with synthetic catalysts. Since then, asymmetric catalysis has become a very active, rapidly growing area of research. As a result, a number of powerful synthetic catalysts have become available which allow the preparation of chiral compounds in high enantiomeric purity [4]. However, despite impressive achievements, the number of truly useful, generally applicable enantioselective catalysts is still limited. Thus, future research will continue to focus on the improvement of existing methods and on the search for new catalysts. In addition, the development of more rational guidelines for the design of enantioselective catalysts, which are based on structural and mechanistic considerations rather than intuition or purely empirical rules, will become more and more important.

Most of the efficient catalysts, developed so far, are metal complexes containing a chiral organic ligand which controls a metal-mediated process in such a way that one of two enantiomeric products is formed with high preference over the other. Some of the requirements for an effective chiral ligand have been summarized by Kagan [5]. The synthesis should be relatively easy and give access to both enantiomers as well as structural variants. The chiral ligand must stay coordinated to the metal during the enantioselectivity-determining step. The catalytic activity of the metal should not drop significantly upon complexation with the chiral ligand, the ideal case being a ligand with an accelerating effect on the catalytic reaction (ligand-accelerated catalysis [6]). An important aspect that has to be considered in the design of a chiral ligand is its conformational flexibility. An ideal ligand, when coordinated to a metal, should adopt a single conformation which allows optimal stereochemical control. For a conformationally flexible ligand, the chances are high that some of the conformations, which are in the equilibrium with each other, will lead to low stereoselectivity.

Stereoselective Synthesis
Editors: Ottow, Schöllkopf, Schulz

2. Chiral C_2-Symmetric Semicorrins

Inspired by the structures of corrinoid and porphinoid metal complexes, which in nature play a fundamental role as biocatalysts, we have developed a route to chiral C_2-symmetric semicorrins **1**, a class of bidentate nitrogen ligands specifically designed for enantioselective catalysis [7]. These ligands possess a number of structural features that make them attractive ligands for the stereocontrol of metal-catalyzed reactions.

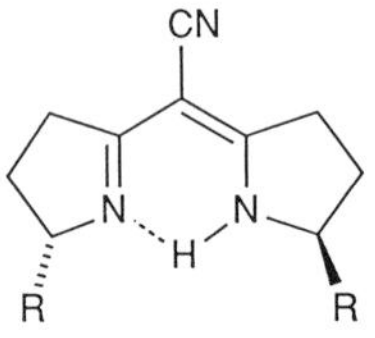

1 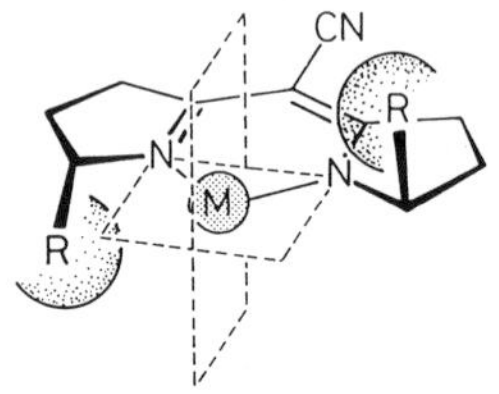 2

First, the geometry of the vinylogous amidine system is ideal for coordinating a metal ion. Accordingly, semicorrins form stable chelate complexes with a variety of metal ions such as Co(II), Rh(I), Ni(II), Pd(II), or Cu(II) [8-10]. Depending on the metal ion, the ligand structure, and the reaction conditions, either mono- or bis(semicorrinato) complexes are obtained. The planar π-system and the two five-membered rings confine the conformational flexibility of the ligand framework. This simplifies the problem of predicting or analyzing the three-dimensional structure of semicorrin metal complexes. Moreover, the conformational rigidity of the ligand system and its C_2-symmetry should have a favorable effect on the stereoselectivity of a metal-catalyzed process, as they restrict the number of possible catalyst-substrate arrangements and the number of competing diastereomeric transition states [11]. The two substituents at the stereogenic centers are located in close proximity to the coordination site. As shown in Formula **2**, they shield the metal ion from two opposite directions and, therefore, are expected to have a distinct, direct effect on a reaction taking place in the coordination sphere. All these properties facilitate an analysis of the possible interactions between catalyst and substrate which determine the stereoselectivity of a metal-catalyzed process and should make it easier to approach the problem of developing an enantioselective catalyst in a straightforward, rational manner.

Semicorrins have been previously prepared as intermediates in the synthesis of corrinoid and hydroporphinoid compounds [12, 13]. The classic route by iminoester-enamine condensation, devised by Eschenmoser [12], is ideally suited for preparing chiral C_2-symmetric semicorrins **1** (Scheme 1). Starting either from L-pyroglutamic acid (–)-**3** or from the D-enantiomer, which are both commercially available at moderate prices, the crystalline, enantiomerically pure (*S,S*)- and (*R,R*)-diesters (–)-**1a** and (+)-**1a** are readily synthesized in multigram quantities with an overall yield of 30-40 % (Scheme 2) [8, 14]. The diesters can be converted to a variety of differently substituted semicorrins by selective trans-

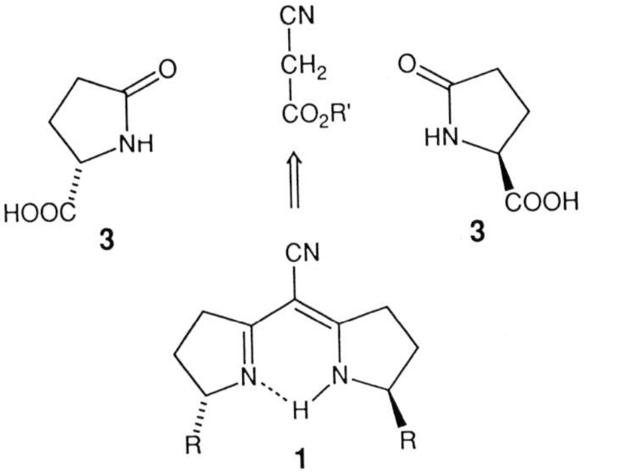

Scheme 1

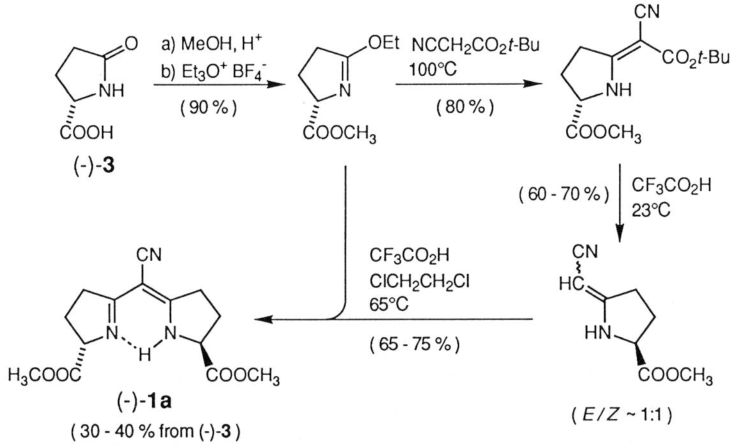

Scheme 2

formation of the ester groups, such as reduction and subsequent silylation leading to **1b** or Grignard reaction producing the diol **1c** (Scheme 3) [8, 15]. Alternatively, the substituents at the stereogenic centers may be altered at the first stage of the synthesis by modifying the carboxyl group of pyroglutamic acid [7a, 15]. This allows the ligand structure to be adjusted to the specific requirements of a particular application and provides a means for optimizing the selectivity of a catalyst in a systematic manner.

Chiral semicorrins have proven to be effective ligands for the enantioselective Cu-catalyzed cyclopropanation of olefins with diazo compounds [16] and for the enantioselective Co-catalyzed reduction of electrophilic C=C double bonds with sodium borohydride [7]. In the presence of 1 mol% of catalyst, generated *in situ* from the Cu(II) complex **4** or from the free ligand **1c** and copper(I) *tert*-butoxide, terminal and 1,2-disubstituted olefins such as styrene, butadiene, or (*E*)-1-phenyl-1-propene react with diazoacetates to give the corresponding cyclopropanecarboxylates with high enantiomeric excess (Scheme 4) [17]. Good

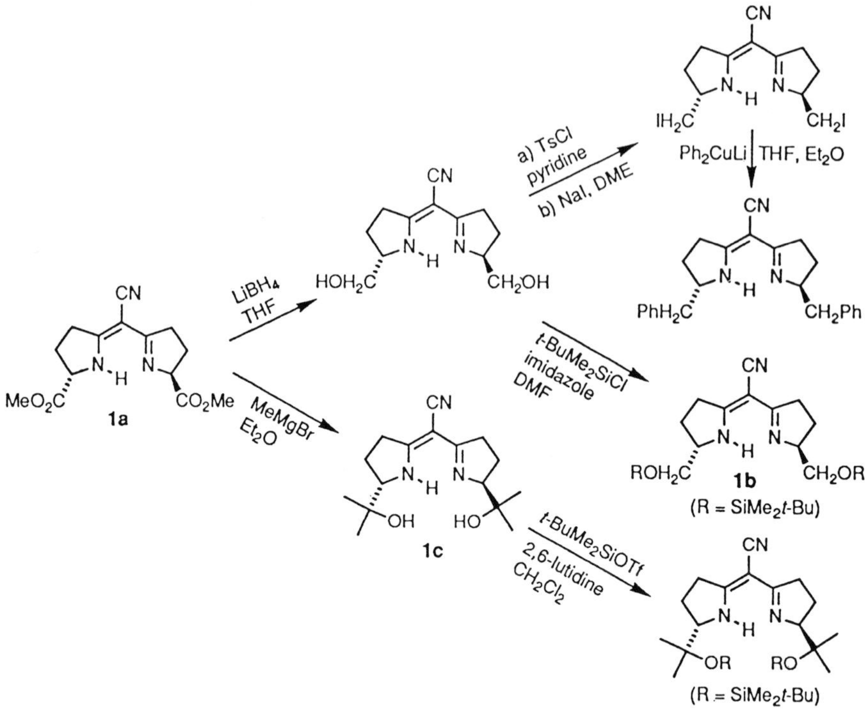

Scheme 3

results have also been obtained with enol ethers as substrates and with intramolecular cyclopropanation reactions of alkenyl diazoketones (Scheme 5) [18].

Cobalt complexes, prepared from cobalt(II) chloride and ligand **1b**, are highly efficient catalysts for the conjugate reduction of α,β-unsaturated carboxylic esters [19] and carboxamides [20] (Schemes 6 and 7). Using 0.1-1 mol% of catalyst and sodium borohydride as reducing agent, the corresponding saturated esters and amides are formed in high enantiomeric purity and essentially quantitative yield.

3. Analogues of Semicorrins: Aza-semicorrins and Bis (oxazolines)

Because of the negative charge and the electron-rich vinylogous amidine system, semicorrins are expected to act as σ- and π-electron-donors which distinctly reduce the electrophilicity of a coordinated metal ion. However, for certain applications, neutral ligands which are weaker electron-donors, or even π-acceptors would seem to be better suited. Therefore, we have prepared various structural analogues of semicorrins **1**, such as the neutral ligands **7** which are readily available from aza-semicorrins **5** by *N*-methylation, the bioxazolines **8**, and the

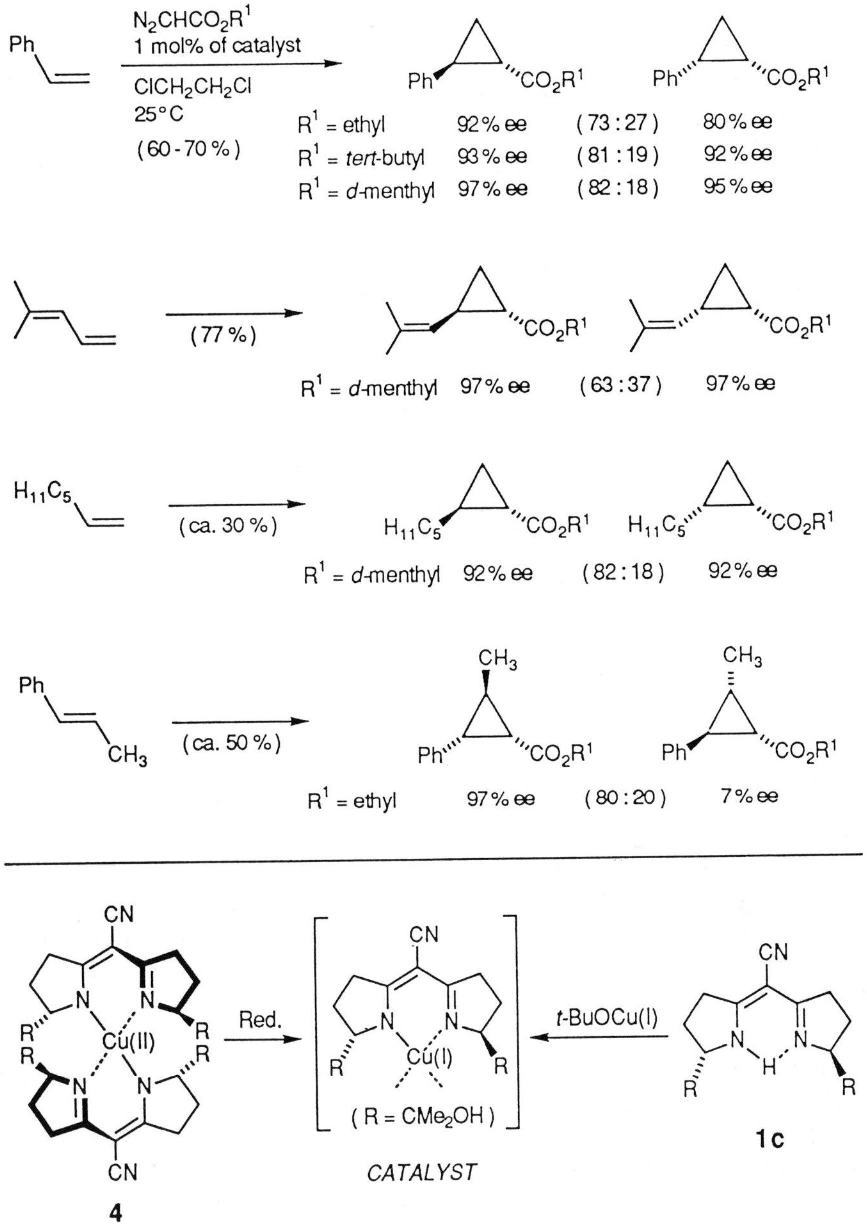

Ph

$N_2CHCO_2R^1$
1 mol% of catalyst
$ClCH_2CH_2Cl$
25°C
(60-70%)

Ph CO_2R^1 Ph CO_2R^1

R^1 = ethyl	92% ee	(73:27)	80% ee
R^1 = *tert*-butyl	93% ee	(81:19)	92% ee
R^1 = *d*-menthyl	97% ee	(82:18)	95% ee

(77%)

CO_2R^1 CO_2R^1

R^1 = *d*-menthyl 97% ee (63:37) 97% ee

$H_{11}C_5$

(ca. 30%)

$H_{11}C_5$ CO_2R^1 $H_{11}C_5$ CO_2R^1

R^1 = *d*-menthyl 92% ee (82:18) 92% ee

Ph

CH_3

(ca. 50%)

CH_3 CH_3

Ph CO_2R^1 Ph CO_2R^1

R^1 = ethyl 97% ee (80:20) 7% ee

CN

N N

R R

Cu(II)

R R

N N

CN

4

Red.

CN

N N

R Cu(I) R

(R = CMe_2OH)

CATALYST

t-BuOCu(I)

CN

N N

H

R R

1c

Scheme 4

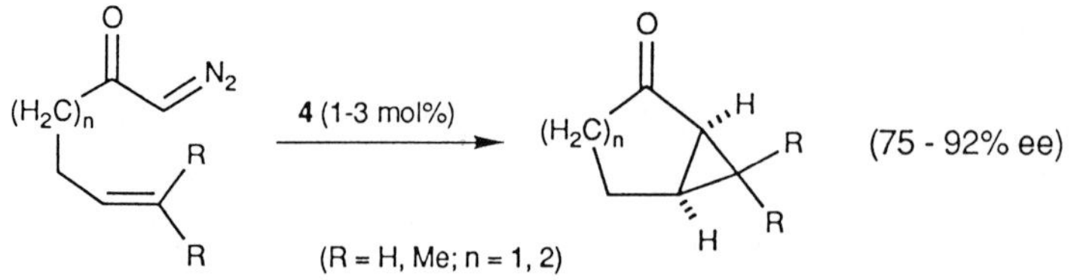
O
N_2
$(H_2C)_n$
R
R
4 (1-3 mol%)
(R = H, Me; n = 1, 2)
O
$(H_2C)_n$
H
R
R
H
(75 - 92% ee)

Scheme 5

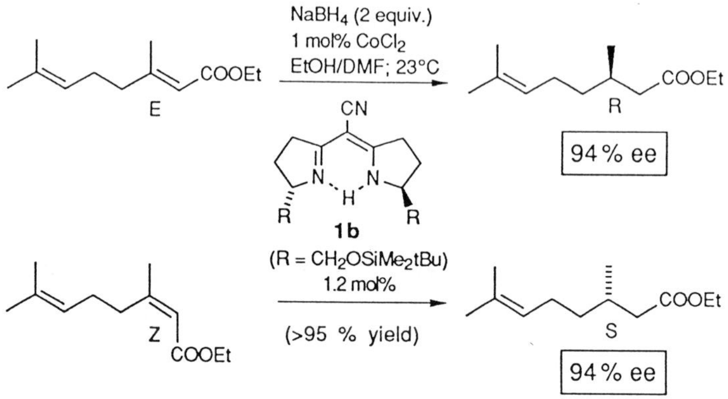
COOEt
E
$NaBH_4$ (2 equiv.)
1 mol% $CoCl_2$
EtOH/DMF; 23°C
CN
N
H
N
R
R
1b
(R = CH_2OSiMe_2tBu)
1.2 mol%
(>95 % yield)
Z
COOEt
COOEt
R
94 % ee
COOEt
S
94 % ee

Scheme 6

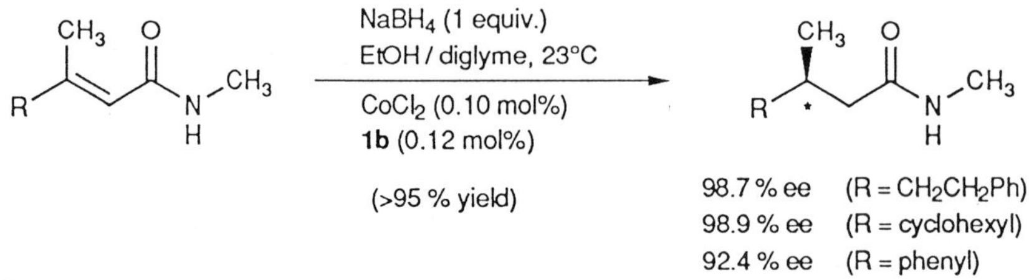
CH_3 O
R
N
H
CH_3
$NaBH_4$ (1 equiv.)
EtOH / diglyme, 23°C
$CoCl_2$ (0.10 mol%)
1b (0.12 mol%)
(>95 % yield)
CH_3 O
R
N
H
CH_3
98.7 % ee (R = CH_2CH_2Ph)
98.9 % ee (R = cyclohexyl)
92.4 % ee (R = phenyl)

Scheme 7

methylene-bis(oxazolines) **9** containing two alkyl groups at the methylene bridge which prevent deprotonation at this position (Scheme 8).

Methylene-bis(oxazolines) **6**, derived from malonate, have a coordination sphere which closely resembles that of the semicorrins **1**. Removal of a proton at the methylene bridge leads to anionic ligands which provide essentially the same steric environment for a coordinated metal ion as their semicorrin counterparts and, therefore, we thought, could be useful substitutes for semicorrins (Scheme 9).

Chiral bis(oxazolines) such as **6**, **8**, or **9** are attractive ligands because they are easily synthesized from amino alcohols (Scheme 9). A wide variety of enantiomerically pure amino alcohols are commercially available while others are readily prepared by reduction of α-amino acids. Although oxazolines have

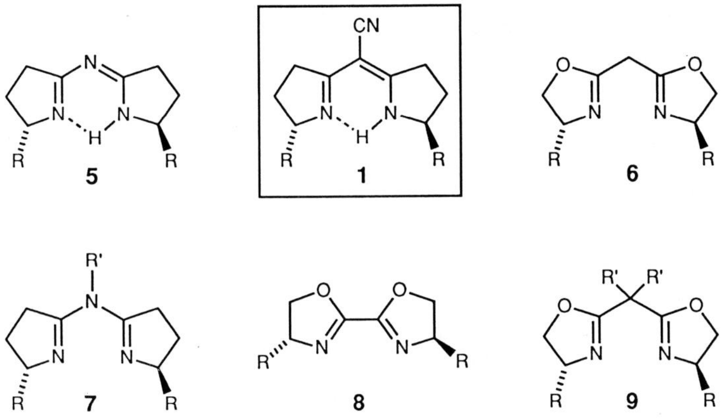

Scheme 8

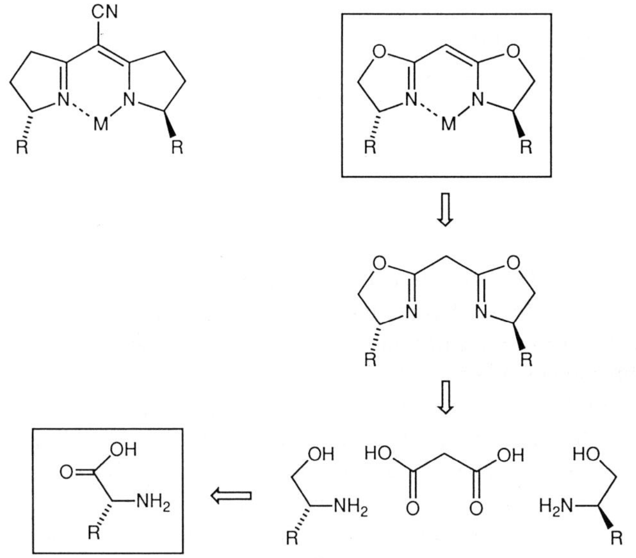

Scheme 9

frequently served as versatile chiral auxiliaries in asymmetric synthesis [21], their potential as ligands for enantioselective catalysis has not been recognized until quite recently. Just after we had started to investigate C_2-symmetric (bis)oxazolines, several applications of chiral oxazolines for the stereocontrol of metal-catalyzed reactions were reported. Bidentate oxazoline ligands derived from pyridine-2-carboxylate were used for the Rh-catalyzed hydrosilylation of ketones (up to 83% ee) and for the Cu-catalyzed monophenylation of *meso*-diols with $Ph_3Bi(OAc)_2$ (up to 50% ee) [22]. Nishiyama *et al.* developed an interesting new class of C_2-symmetric tridentate bis(oxazoline) ligands derived from

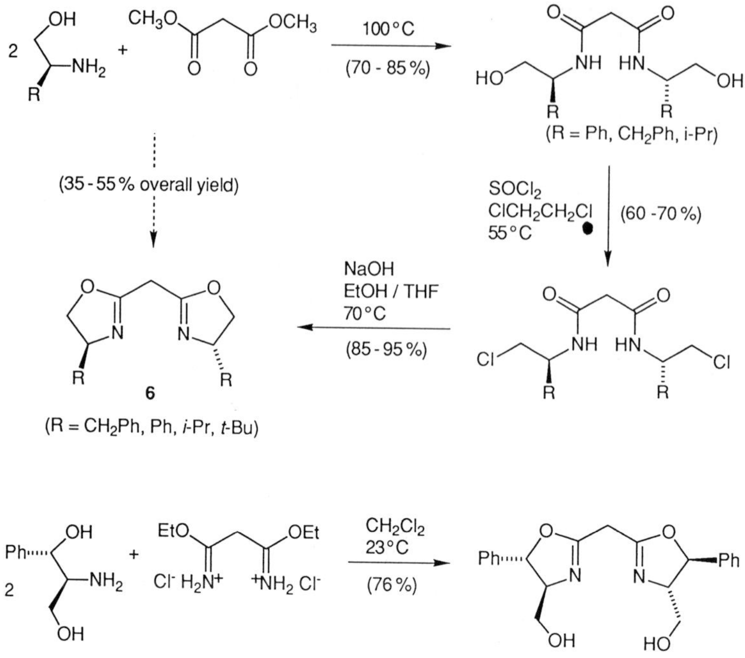

(35-55% overall yield)

(R = Ph, CH_2Ph, i-Pr)

(R = CH_2Ph, Ph, *i*-Pr, *t*-Bu)

Scheme 10

pyridine-2,6-dicarboxylate [23]. Rhodium(III)-complexes of these ligands proved to be efficient catalysts for the hydrosilylation of ketones (up to 99% ee). The synthesis and structure of a series of related oxazoline metal complexes has been described by Bolm and coworkers [24].

The syntheses of chiral C_2-symmetric methylene-bis(oxazolines) and bi-oxazolines of type **6** and **8** are summarized in Schemes 10 and 11 [25]. Using the well-established three-step sequence - amide formation, conversion to the bis(2-chloroalkyl)amide, and subsequent base-induced cyclization [26] - various derivatives were prepared in good overall yields. Alternatively, these ligands may be synthesized in one step from the corresponding amino alcohols and readily available bis-imidates derived from malonate or oxalate [25, 27] or, as described by Masamune *et al.* [28], by amide formation with diethyl malonate and subsequent reaction with dimethyltin dichloride, or by metal-catalyzed condensation of amino alcohols with nitriles [29].

5-Aza-semicorrins are readily assembled from appropriate butyrolactam derivatives (Scheme 12) [30]. As for the semicorrins **1**, pyroglutamic acid or the corresponding alcohol **10**, which are both commercially available, serve as versatile precursors. As expected, methylation of ligand **11** occurs exclusively at the aza-bridge rather than at the nitrogen atoms in the hydropyrrol rings, which are protected by the intramolecular hydrogen bond. The overall yield of the neutral ligand **12**, starting from **10**, is in the range of 20-30%. A much shorter route to this ligand system is based on the method developed by Vorbrüggen for the con-

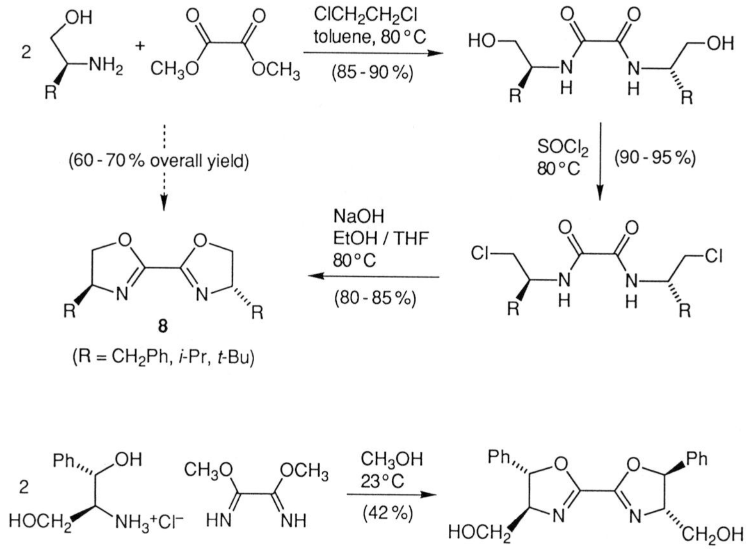

Scheme 11

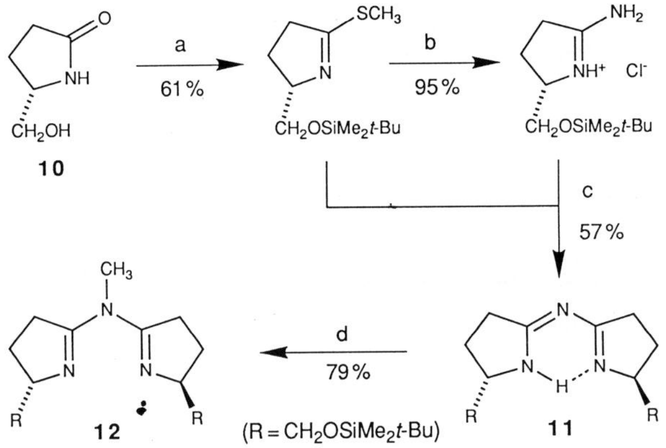

Reagents

(a) Me_2t-BuSiCl, imidazole, 40°C; Lawesson's reagent, THF, 23°C; MeI, CH_2Cl_2, 23°C; $NaHCO_3$, CH_2Cl_2, H_2O.

(b) NH_4Cl, MeOH, reflux

(c) BuLi (0.9 equiv), THF, 0-23°C; evacuated ampoule, 72°C.

(d) MeI, 23°C; $NaHCO_3$, CH_2Cl_2, H_2O.

Scheme 12

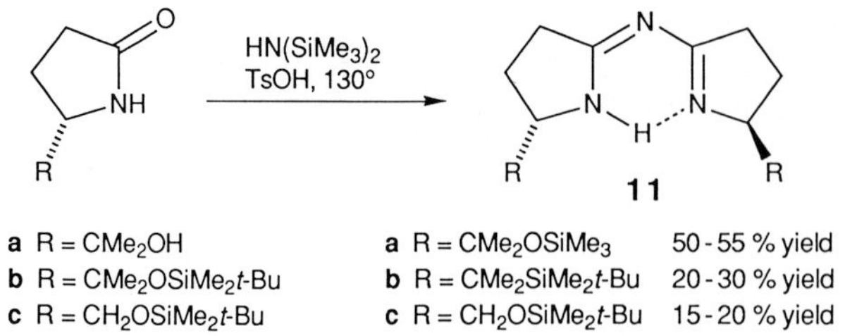

a R = CMe$_2$OH
b R = CMe$_2$OSiMe$_2$t-Bu
c R = CH$_2$OSiMe$_2$t-Bu

a R = CMe$_2$OSiMe$_3$ 50 - 55 % yield
b R = CMe$_2$SiMe$_2$t-Bu 20 - 30 % yield
c R = CH$_2$OSiMe$_2$t-Bu 15 - 20 % yield

Scheme 13

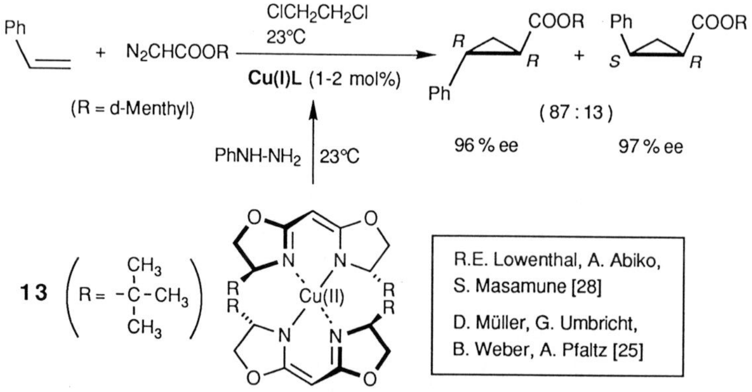

R.E. Lowenthal, A. Abiko, S. Masamune [28]

D. Müller, G. Umbricht, B. Weber, A. Pfaltz [25]

Scheme 14

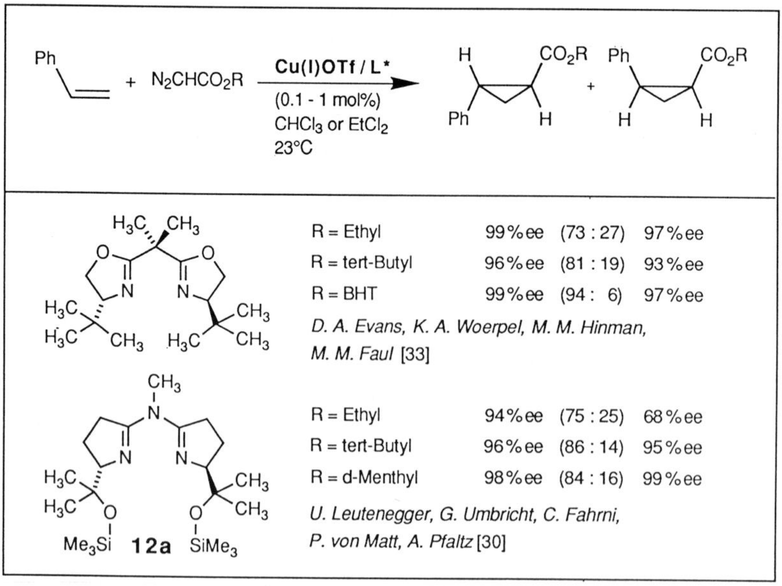

R = Ethyl	99%ee	(73 : 27)	97%ee
R = tert-Butyl	96%ee	(81 : 19)	93%ee
R = BHT	99%ee	(94 : 6)	97%ee

D. A. Evans, K. A. Woerpel, M. M. Hinman, M. M. Faul [33]

R = Ethyl	94%ee	(75 : 25)	68%ee
R = tert-Butyl	96%ee	(86 : 14)	95%ee
R = d-Menthyl	98%ee	(84 : 16)	99%ee

U. Leutenegger, G. Umbricht, C. Fahrni, P. von Matt, A. Pfaltz [30]

Scheme 15

version of amides to amidines (Scheme 13) [31]. Treatment of butyrolactams, derived from pyroglutamic acid, with bis(trimethylsilyl)amine in the presence of a catalytic amount of acid, directly leads to the corresponding aza-semicorrins **11** [30]. The best yields were obtained for **11a**, whereas other derivatives were formed in lower and less reproducible yields.

The obvious first test of methylene-bis(oxazolines) **6** was the copper-catalyzed cyclopropanation. Among the various derivatives that we had prepared (**6**; R = benzyl, *sec*-alkyl, phenyl, *tert*-butyl), the bulky *tert*-butyl-substituted derivative proved to be the most effective ligand, giving similarly high enantiomeric excesses as the semicorrin **1c** (R = CMe_2OH) (Scheme 14) [25]. The same results were obtained by Masamune and coworkers in an independent parallel study [28]. More recently, Lowenthal and Masamune showed that the method works also well for certain trisubstituted and unsymmetrically disubstituted cis-olefins if the di-*tert*-butyl-substituted ligand is replaced by structurally modified bis(oxazolines) [32].

A related but even more selective, highly efficient catalyst, prepared from a neutral (bis)oxazoline ligand and Cu(I) triflate, was described by Evans *et al.* (Scheme 15) [33]. Terminal olefins such as styrene and 2-methylpropene were converted to cyclopropane-carboxylates with up to 99% ee, using substrate-catalyst ratios as high as 1000:1. The same catalyst was used for an analogous enantioselective aziridine formation. In view of these results, we briefly investigated analogous catalysts derived from neutral 5-aza-semicorrins and Cu(I) triflate. The best results were obtained with sterically demanding ligands such as **12a**. Taking the enantiomeric purity of the trans-product obtained from ethyl diazoacetate and styrene as a standard, the selectivities of the aza-semicorrin catalysts ranged in between those of the Evans catalyst and the semicorrin and (bis)oxazoline copper complexes **4** and **13**.

The potential of C_2-symmetric bis(oxazolines) as ligands in asymmetric catalysis has also been recognized by other groups (Scheme 16) [34]. Corey *et al.* reported an example of an enantioselective Diels-Alder reaction catalyzed by an iron(III) complex with a neutral methylene-bis(oxazoline) ligand [35]. Helmchen and coworkers prepared a series of differently substituted bioxazolines and analogous bithiazolines which they employed in the Rhodium-catalyzed enantioselective hydrosilylation of acetophenone [36]. The Lehn group described the synthesis and X-ray analysis of the crystalline Cu(II) (bis)oxazoline complex **14** [37]. The crystal structures of **14** and the analogous bis(semicorrinato) complex **4** [8b] demonstrate the close structural resemblance of semicorrin and methylene-(bis)oxazoline ligands. The preparation of Rh(I), Ru(II), Cu(I), and Pd(II) complexes of bioxazolines was reported by Onishi and Isagawa [38].

Originally, we had hoped that transition metal complexes with neutral oxazoline ligands could be used as hydrogenation catalysts. Although screening of various Rh, Ru, and Ir complexes did not reveal any apparent reactivity towards molecular hydrogen, Ir(I) complexes prepared *in situ* from $[Ir(COD)Cl]_2$ were found to catalyze the reduction of ketones using isopropanol as hydride donor (Scheme 17) [25]. The best selectivity was observed in the reduction of isopropyl phenyl ketone. These results compare favorably to the selectivities report-

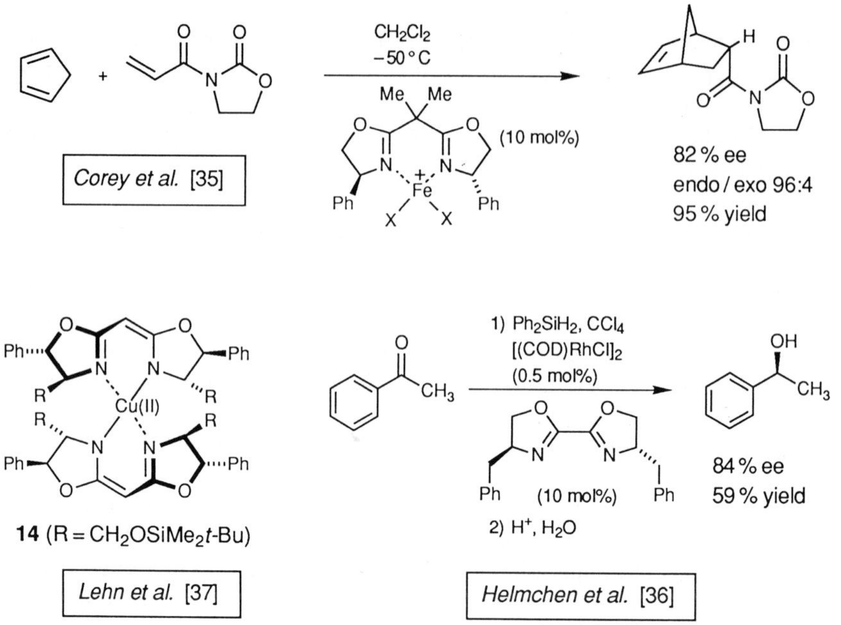

CH_2Cl_2
−50 °C
(10 mol%)
82 % ee
endo / exo 96:4
95 % yield

Corey et al. [35]

14 (R = CH_2OSiMe_2*t*-Bu)

Lehn et al. [37]

1) Ph_2SiH_2, CCl_4
$[(COD)RhCl]_2$
(0.5 mol%)
(10 mol%)
2) H^+, H_2O
84 % ee
59 % yield

Helmchen et al. [36]

Scheme 16

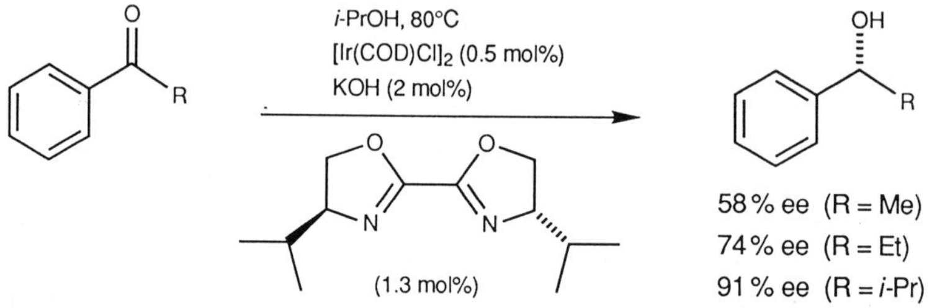

i-PrOH, 80°C
$[Ir(COD)Cl]_2$ (0.5 mol%)
KOH (2 mol%)
(1.3 mol%)

58 % ee (R = Me)
74 % ee (R = Et)
91 % ee (R = *i*-Pr)

Scheme 17

ed for other iridium catalysts [39]. Dialkyl ketones proved to be unreactive under these conditions. Ir complexes with neutral methylene-bis(oxazoline) ligands of type **9** or with anionic semicorrin or bis(oxazoline) ligands did not exhibit any significant catalytic activity.

All these results point to a considerable potential of semicorrins and analogous bis(oxazolines) and aza-semicorrins in asymmetric catalysis. Starting from commercially available amino alcohols or amino acids, a wide variety of such ligands can be prepared in enantiomerically pure form. By proper selection of the ligand structure, the steric influence, the coordination geometry, and the electronic properties of the ligand the catalyst can be adjusted to the specific require-

$Na^+ {}^-CH(CO_2Me)_2$, THF, 50°C	86 % yield	77 % ee (R)
$Na^+ {}^-CH(CO_2Me)_2$, THF, 50°C	85 % yield	76 % ee (R)
$CH_2(CO_2Me)_2$, BSA*), KOAc, THF, 50°C	95 % yield	73 % ee (R)
$CH_2(CO_2Me)_2$, BSA*), KOAc, CH_2Cl_2, 23°C	97 % yield	88 % ee (R)
$CH_2(CO_2Me)_2$, BSA*), KOAc, CH_2Cl_2, 23°C	99 % yield	95 % ee (R)
$CH_2(CO_2Me)_2$, BSA*), KOAc, CH_2Cl_2, 4°C	99 % yield	95 % ee (R)

*) BSA = $H_3C-C(O-SiMe_3)=N-SiMe_3$

Scheme 18

ments of a particular application. The ready access to these ligands and the ease of modifying their structures offer excellent opportunities for the development of taylor-made enantioselective catalysts for many other classes of reactions.

4. Pd-Catalyzed Allylic Substitution

Thanks to the fundamental studies of Tsuji, Trost, and others, palladium-catalyzed allylic substitution has become a versatile, widely used process in organic synthesis [40]. The search for efficient enantioselective catalysts for this class of reactions is an important goal of current research in this field [41]. It has been shown that chiral phosphine ligands can induce substantial enantiomeric excesses in Pd-catalyzed reactions of racemic or achiral allylic substrates with nucleophiles [42]. Recently, promising results have also been obtained with chiral bidentate nitrogen ligands [43]. We have found that palladium complexes of neutral aza-semicorrin or methylene-bis(oxazoline) ligands are effective catalysts for the enantioselective allylic alkylation of 1,3-diphenyl-2-propenyl acetate or related substrates with dimethyl malonate (Schemes 18 [25, 30] and 19 [44]).

The best results were obtained in apolar media using a mixture of dimethyl malonate and *N,O*-bis(trimethylsilyl)acetamide (BSA), according to a procedure

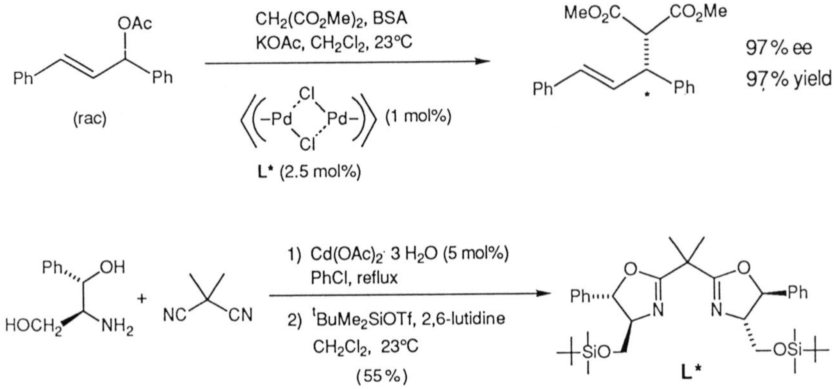

Scheme 19

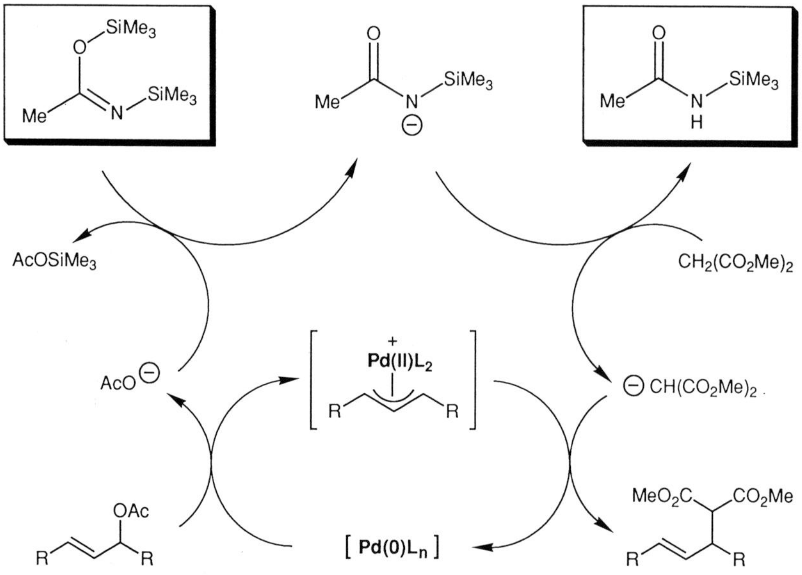

Scheme 20

described by Trost [45]. The catalytic process was initiated by addition of a catalytic amount of potassium acetate (Scheme 20). Under these conditions, in the presence of 1-2 mol% of catalyst, the reaction proceeded smoothly at room temperature to give the desired product in high enantiomeric purity and essentially quantitative yield.

How can we explain the remarkable enantioselectivities induced by these ligands? According to the generally accepted mechanism of palladium-catalyzed allylic alkylation [40, 41], the reaction is assumed to proceed by the pathway shown in Scheme 21. Because the two substituents at the allylic termini are identical, the same (allyl)Pd(II) intermediate **15** is formed starting from the two enantiomers of the racemic substrate. The enantioselectivity of the overall reaction is determined by the regioselectivity of the subsequent step involving nucleophilic cleavage of one of the (Pd-C) bonds, a type of process which has been shown to take place with inversion of configuration [41, 42b, 43a]. Under the influence of the chiral ligand, the nucleophile attacks preferentially one of the two allylic termini. So the question is, which factors determine the regioselectivity of this step?

Some clues are provided by the crystal structures of the two (allyl)Pd complexes shown in Figure 1 [46]. The complex **18a** with the unsubstituted allyl li-

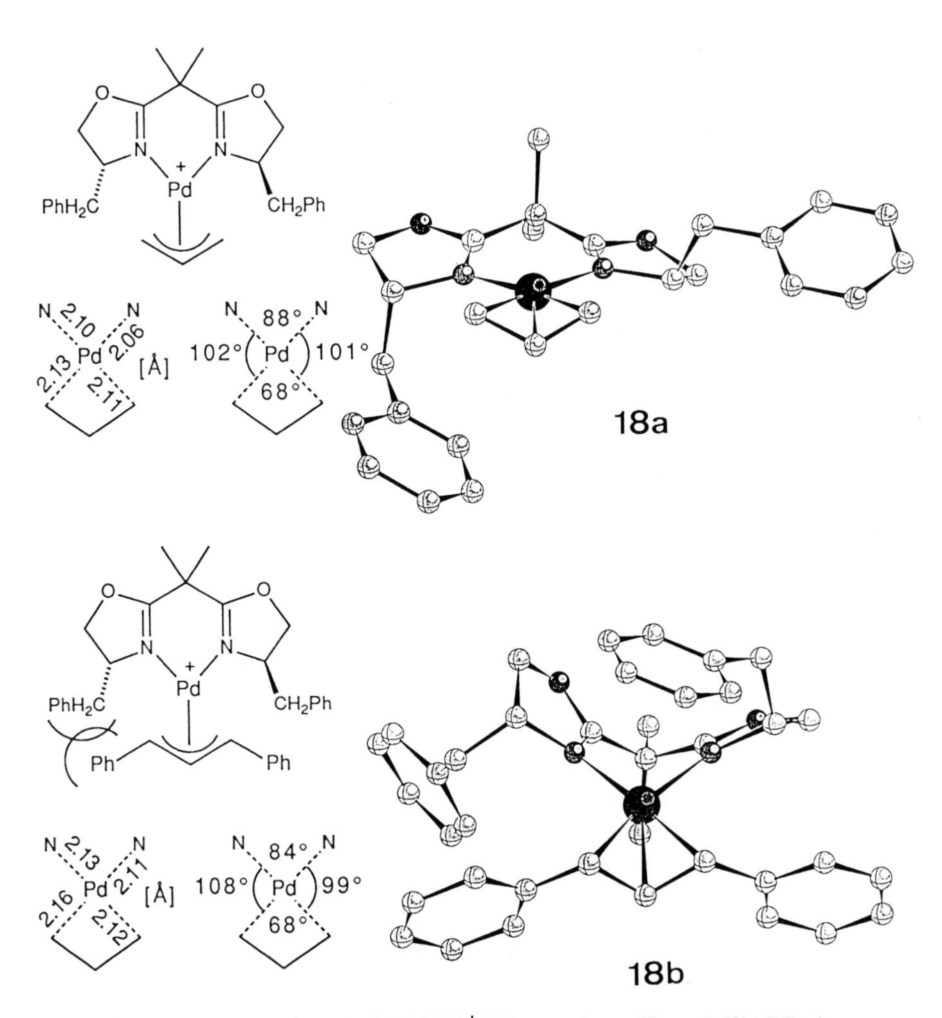

Fig. 1. Crystal structures of the $[(allyl)Pd(II)]^+$ PF_6^- complexes **18a** and **18b** (PF_6^- ions not shown) [46].

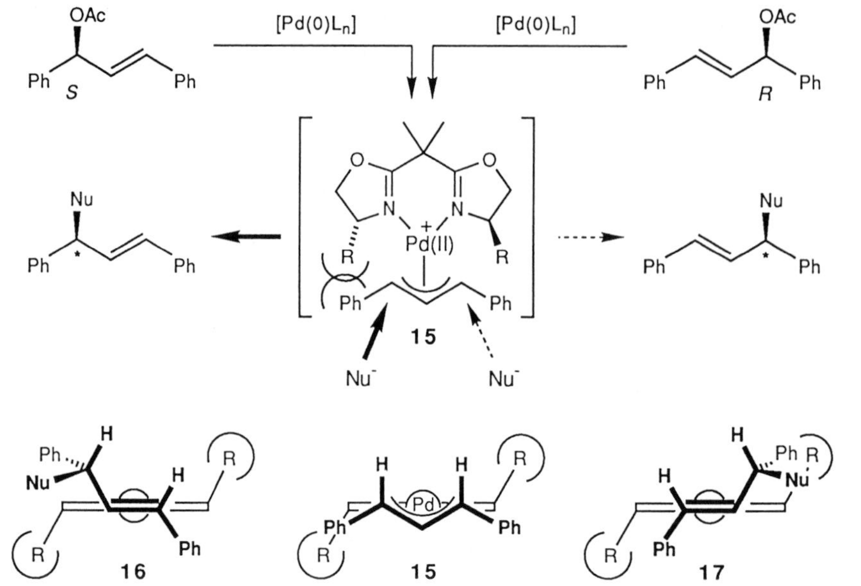

Scheme 21

gand shows the expected square planar coordination geometry of Pd(II) and an almost planar conformation of the methylene-bis(oxazoline) ligand framework. The structure of the corresponding (1,3-diphenylallyl)Pd complex **18b**, which is the actual intermediate in the catalytic reaction (cf. Scheme 21), is strikingly different. As a consequence of the steric repulsion between the allylic phenyl group and the adjacent benzyl substituent of the chiral ligand, the methylene-bis(oxazoline) ring system adopts a strongly distorted, non-planar conformation. The repulsive interaction between the chiral ligand and one of the allylic termini is also reflected in the bond lengths and bond angles of the $[PdC_2N_2]$ core (see Figure 1). From the absolute configuration of the product we know that the nucleophile preferentially attacks the longer, more strained (Pd–C) bond (see Scheme 21), suggesting that the release of steric strain, associated with the cleavage of this bond, may be one of the factors responsible for enantioselection. This concept of regioselective steric activation of one of the (Pd-C) bonds differs from previously developed concepts which are based on direct interactions between the chiral ligand and the nucleophile [42a, c, 47]. Another possible factor contributing to the observed selectivity may be related to the thermodynamic stability of the resulting Pd(0)-olefin complexes which are postulated as primary products [41]. Assuming a square-planar coordination geometry [48], the Pd(0) complex **16** of the major product enantiomer is expected to be more stable than **17**, based on steric considerations (see Scheme 21). Analogous steric interactions between the chiral ligand and the coordinated substrate should also be present in the corresponding transition states.

Palladium catalysts with aza-semicorrin and bis(oxazoline) ligands have also been successfully applied in analogous reactions of 1-aryl-3-alkyl- and 1-

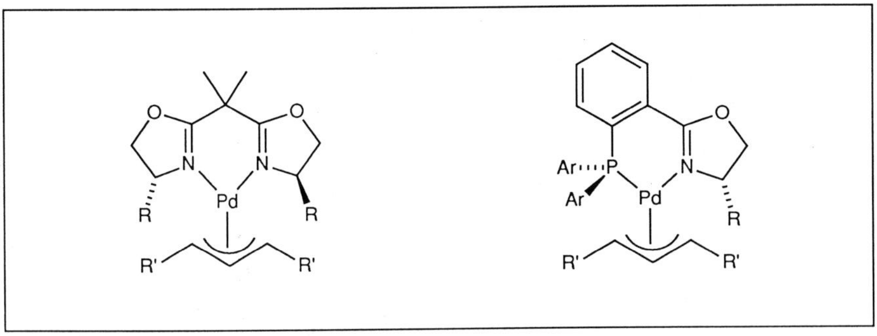
O
N
N
O
Pd
R
R
R'
R'
Ar
Ar
P
Pd
N
O
R
R'
R'

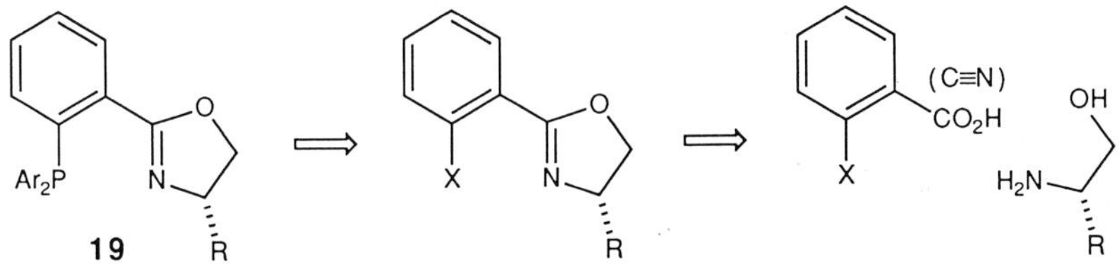
Ar_2P
N
O
19
R
X
N
O
R
(C≡N)
CO_2H
X
OH
H_2N
R

Scheme 22

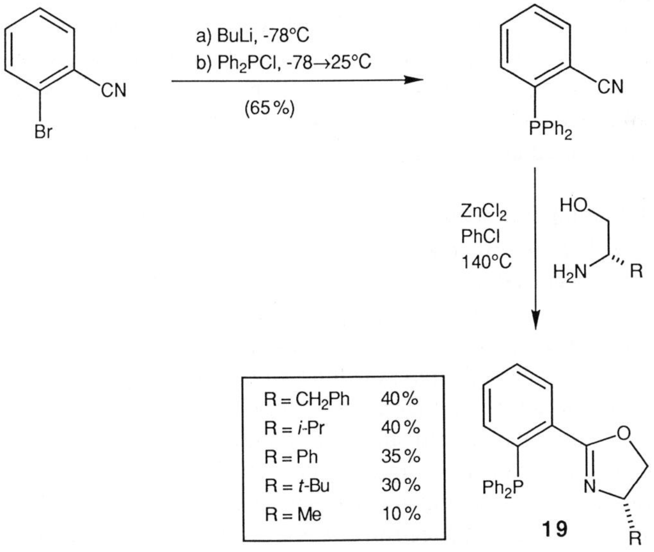
CN
Br
a) BuLi, -78°C
b) Ph_2PCl, -78→25°C
(65%)
CN
PPh_2
$ZnCl_2$
PhCl
140°C
HO
H_2N
R
R = CH_2Ph 40%
R = *i*-Pr 40%
R = Ph 35%
R = *t*-Bu 30%
R = Me 10%
Ph_2P
N
O
19
R

Scheme 23

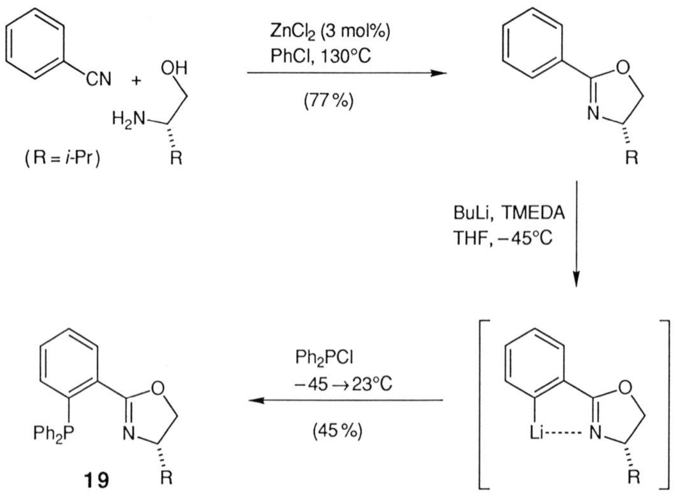

Scheme 24

alkinyl-3-aryl-substituted allylic acetates [49]. 1,3-Dialkyl-2-propenyl acetates, on the other hand, react very sluggishly and with poor enantioselectivity. We have now developed a new class of ligands, chiral 2-(2-phosphinoaryl)-2-oxazolines (Scheme 22), which overcome this limitation [50].

2-(2-Phosphinophenyl)-oxazolines [51] such as **19** can be easily prepared in two steps from 2-bromobenzonitrile (Scheme 23). An alternative route involves oxazoline formation from benzonitrile and an amino alcohol, followed by ortho-metallation with BuLi [52] and subsequent reaction with Ph_2PCl (Scheme 24). Starting from commercially available optically active amino alcohols, a variety of differently substituted enantiomerically pure ligands is readily accessible.

Palladium(II)-allyl complexes with these ligands were found to be effective catalysts for allylic alkylations with stabilized carbanions. Under standard conditions, using 2 mol% of catalyst and a mixture of BSA and catalytic amounts of KOAc as base, racemic 1,3-diphenyl-2-propenyl acetate smoothly reacted with dimethyl malonate or acetylacetone to afford optically active substitution products (Table 1). The phosphinophenyl-oxazoline with a phenyl group at the stereogenic center proved to be the optimal ligand for this substrate. After a remarkably short reaction time, the desired substitution products were isolated in essentially quantitative yields with enantiomeric excess of 97-99%. These values exceed the selectivities previously obtained with other ligands [41-43].

Table 2 summarizes the results obtained with 1,3-dialkyl-2-propenyl acetates - a class of substrates for which enantioselective allylic alkylations have not been reported yet [53]. Here, the ligand with a *tert*-butyl group at the stereogenic center gave the best enantioselectivities. The 1,3-dimethyl and 1,3-di-*n*-propyl-allyl derivatives react with dimethyl malonate under mild conditions to afford the expected products in high yields with ~70% ee. The 1,3-diisopropyl derivative was

Table 1. Enantioselective allylic alkylation of 1,3-diphenyl-2-propenyl acetate

Ph–CH=CH–CH(OAc)–Ph + H-Nu → Ph–CH=CH–C*H(Nu)–Ph

1 mol% $[Pd(C_3H_5)Cl]_2$ / 2.5 mol% L*; BSA, 2 mol% KOAc, CH_2Cl_2, 23 °C (>95% yield)

Ligand (Ph₂P-phenyl-oxazoline, R)	Nucleophile: MeOOC–CH⁻–COOMe	Nucleophile: MeCO–CH⁻–COMe	Nucleophile: MeCO–C⁻(NHAc)–COMe
R = Me	89% *ee*	84% *ee*	77% *ee*
R = CH2Ph	97% *ee*	93% *ee*	91% *ee*
R = *i*-Pr	98% *ee*	92% *ee*	96% *ee*
R = Ph	99% *ee*	97% *ee*	97% *ee*
R = *t*-Bu	95% *ee*	94% *ee*	96% *ee*

Table 2. Enantioselective allylic alkylation of 1,3-dialkyl-2-propenyl acetates

R–CH=CH–CH(OAc)–R (*rac*) + ⁻CH(COOMe)₂ → R–CH=CH–C*H(R)–CH(COOMe)₂

1 mol% $[Pd(C_3H_5)Cl]_2$, 2.5 mol% L*

a: R = Me	$CH_2(COOMe)_2$, BSA, CH_2Cl_2, 23°C, 96 h	71% ee (96% yield)
b: R = ⁿPr	$CH_2(COOMe)_2$, BSA, CH_2Cl_2, 23°C, 96 h	69% ee (96% yield)
c: R = ⁱPr	$NaCH(COOMe)_2$, DMF, 65°C, 40 h	96% ee (88% yield)

L* = (Ph₂P-phenyl-oxazoline, R = *t*-Bu)

found to be much less reactive (<20% conversion under these conditions). However, under more vigorous conditions with $NaCH(COOMe)_2$ in DMF at 65°C, the reaction proceeds in good yield and with high enantioselectivity.

At present, no simple mechanistic rationale for the observed enantioselection can be offered. Compared to analogous reactions using C_2-symmetric nitrogen ligands, which may be rationalized in a straightforward manner (see above), the situation here is more complex. With non-symmetric phosphi-

noaryl-oxazolines, two diastereomeric allyl palladium intermediates must be considered, and in addition to steric interactions, electronic effects may play a role as well (different *trans*-influence of the coordinating N- and P-atom; see, e.g., ref. [40d]).
Further studies, exploring the scope of these catalysts and the mechanism of enantioselection, are in progress. Considering the promising results obtained so far, it will be interesting to test phosphinoaryl-oxazolines as ligands in other metal-catalyzed reactions.

Acknowledgement. I would like to thank my present and former collaborators, whose names appear in the references, for their many contributions and for their careful experimental work, their enthusiasm, and perseverance. Financial support by the Swiss National Science Foundation and F. Hoffmann-La Roche AG, Basel, is gratefully acknowledged.

References

[1] W. S. Knowles, *Acc. Chem. Res.* **1983**, *16*, 106; K. E. Koenig, in ref. [4a], pp. 71–101.
[2] U. Eder, G. Sauer, R. Wiechert, *Angew. Chem.* **1971**, *83*, 492; *Angew. Chem., Int. Ed. Engl.* **1971**, *10*, 496.
[3] Z. G. Hajos, D. R. Parrish, *J. Org. Chem.* **1974**, *39*, 1615.
[4] (a) *Asymmetric Synthesis*; Vol. 5: *Chiral Catalysis* (Ed.: J. D. Morrison), Academic Press, Orlando, FL, **1985**; (b) *Asymmetric Catalysis* (Ed.: B. Bosnich), NATO ASI Series E 103, Martinus Nijhoff, Dordrecht, **1986**; (c) H. Brunner, *Topics Stereochem.* **1988**, *18*, 129; *Synthesis* **1988**, 645; (d) R. Noyori, M. Kitamura, in *Modern Synthetic Methods 1989* (Ed.: R. Scheffold), Springer, Berlin - Heidelberg, **1989**, pp. 115-198.
[5] H. B. Kagan, in ref. [4a], pp. 1-39.
[6] E. N. Jacobsen, I. Markó, W. S. Mungall, G. Schröder, K. B. Sharpless, *J. Am. Chem. Soc.* **1988**, *110*, 1968.
[7] Reviews: (a) A. Pfaltz, in *Modern Synthetic Methods 1989* (Ed.: R. Scheffold), Springer: Berlin-Heidelberg, **1989**, pp. 199-248; (b) A. Pfaltz, *Chimia* **1990**, *44*, 202; c) A. Pfaltz, *Acc. Chem. Res.* **1993**, *26*, in press.
[8] (a) H. Fritschi, U. Leutenegger, A. Pfaltz, *Angew. Chem.* **1986**, *98*, 1028; *Angew. Chem., Int. Ed. Engl.* **1986**, *25*, 1005; (b) H. Fritschi, U. Leutenegger, K. Siegmann, A. Pfaltz, W. Keller, C. Kratky, *Helv. Chim. Acta* **1988**, *71*, 1541.
[9] H. Fritschi, *Dissertation*, ETH-Zürich, Nr. 8951, **1989**.
[10] A. P. Johnson, P. Wehrli, R. Fletcher, A. Eschenmoser, *Angew. Chem.* **1968**, *80*, 622; *Angew. Chem., Int. Ed. Engl.* **1968**, *7*, 623.
[11] H. B. Kagan, T.-P. Dang, *J. Am. Chem. Soc.* **1972**, *94*, 6429; see also ref. [5]. For a review of C_2-symmetric auxiliaries and ligands, see J. K. Whitesell, *Chem. Rev.* **1989**, *89*, 1581.
[12] A. Eschenmoser, *Quart. Rev. Chem. Soc.* **1970**, *24*, 366; Y. Yamada, D. Miljkovic, P. Wehrli, B. Golding, P. Löliger, R. Keese, K. Müller, A. Eschenmoser, *Angew. Chem.* **1969**, *81*, 301; *Angew. Chem., Int. Ed. Engl.* **1969**, *8*, 343; A. Pfaltz, N. Bühler, R. Neier, K. Hirai, A. Eschenmoser, *Helv. Chim. Acta* **1977**, *60*, 2653.
[13] R. V. Stevens, *Tetrahedron* **1976**, *32*, 1599.
[14] Semicorrins (–)-**1a** and (–)-**1b** are commercially available (FLUKA Chemie AG).
[15] U. Leutenegger, *Dissertation*, ETH-Zürich Nr. 9091, **1990**.
[16] For a recent review of enantioselective metal-catalyzed cyclopropanation, see: M. P. Doyle, *Recl. Trav. Chim. Pays-Bas* **1991**, *110*, 305.
[17] H. Fritschi, U. Leutenegger, A. Pfaltz, *Helv. Chim. Acta* **1988**, *71*, 1553; see also ref. [7], [8a], [9].

[18] C. Piqué, *Dissertation*, University of Basel, **1993**, in press; G. Umbricht, *Dissertation*, University of Basel, **1993**, in press.
[19] U. Leutenegger, A. Madin, A. Pfaltz, *Angew. Chem.* **1989**, *101*, 61; *Angew. Chem., Int. Ed. Engl.* **1989**, *28*, 60.
[20] P. von Matt, A. Pfaltz, *Tetrahedron: Asymmetry* **1991**, *2*, 691.
[21] K. A. Lutomski, A. I. Meyers, in *Asymmetric Synthesis,* Vol. 3 (Ed.: J. D. Morrison), Academic Press, Orlando, FL, **1983**, pp. 213-274.
[22] (a) H. Brunner, U. Obermann, *Chem. Ber.* **1989**, *122*, 499; H. Brunner, P. Brandl, *J. Organomet. Chem.* ***1989***, *390*, C81; G. Balavoine, J. Clinet, I. Lellouche, *Tetrahedron Lett.* ***1989***, *30*, 5141;
(b) H. Brunner, U. Obermann, P. Wimmer, *Organometallics* ***1989***, *8*, 821.
[23] H. Nishiyama, H. Sakaguchi, T. Nakamura, M. Horihata, M. Kondo, K. Itoh, *Organometallics* **1989**, *8*, 846; H. Nishiyama, M. Kondo, T. Nakamura, K. Itoh, *ibid.* **1991**, *10,* 500; H. Nishiyama, S. Yamaguchi, M. Kondo, K. Itoh, *J. Org. Chem.* **1992**, *57*, 4306.
[24] C. Bolm, K. Weickhardt, M. Zehnder, T. Ranff, *Chem. Ber.* **1991**, *124*, 1173; C. Bolm, K. Weickhardt, M. Zehnder, D. Glasmacher, D. *Helv. Chim. Acta* ***1991***, *74*, 717.
[25] D. Müller, G. Umbricht, B. Weber, A. Pfaltz, *Helv. Chim. Acta* **1991**, *74*, 232.
[26] H. Wenker, *J. Am. Chem. Soc.* **1938**, *60*, 2152; I. Butula, G. Karlović, *Liebigs Ann. Chem.* **1976**, 1455; J. A. Frump, *Chem. Rev.* **1971**, *71*, 483.
[27] (a) S. M. McElvain, J. P. Schroeder, *J. Am. Chem. Soc.* **1949**, *71*, 40; A. Meyers, G. Knaus, K. Kamata, M. E. Ford, *ibid.* **1976**, *98*, 567; (b) H. Weidinger, J. Kranz, *Chem. Ber.* **1964**, *97*, 1599.
[28] R. E. Lowenthal, A. Abiko, S. Masamune, *Tetrahedron Lett.* **1990**, *31*, 6005.
[29] H. Witte, W. Seeliger, *Liebigs Ann. Chem.* **1974**, 996.
[30] U. Leutenegger, G. Umbricht, C. Fahrni, P. von Matt, A. Pfaltz, *Tetrahedron* **1992**, *48*, 2143.
[31] H. Vorbrüggen (Schering AG), Ger. Offen. 2,256,755 (1974); *Chem. Abstr.* **1974**, *81*, 63641c; H. Vorbrüggen, K. Krolikiewicz, *Liebigs Ann. Chem.* **1976**, 745; *Chem. Ber.* **1984**, *117*, 1523; see also: H. Vorbrüggen, *Adv. Heterocycl. Chem.* **1990**, *49*, 117. We are grateful to Prof. H. Vorbrüggen for his suggestion to apply his method to the synthesis of aza-semicorrins.
[32] R. E. Lowenthal, S. Masamune, *Tetrahedron Lett.* **1991**, *32*, 7373.
[33] D. A. Evans, K. A. Woerpel, M. M. Hinman, M. M. Faul, *J. Am. Chem. Soc.* **1991**, *113*, 726; D. A. Evans, K. A. Woerpel, M. J. Scott, *Angew. Chem.* **1992**, *104*, 439; *Angew. Chem., Int. Ed. Engl.* **1991**, *31*, 430.
[34] For a short review, see: C. Bolm, *Angew. Chem.* **1991**, *103*, 556; *Angew. Chem., Int. Ed. Engl.* **1991**, *30*, 542.
[35] E. J. Corey, N. Imai, H.-Y. Zhang, *J. Am. Chem. Soc.* **1991**, *113*, 728; E. J. Corey, K. Ishihara, *Tetrahedron Lett.* **1992**, *33*, 6807.
[36] G. Helmchen, A. Krotz, K. T. Ganz, D. Hansen, *Synlett* **1991**, 257.
[37] J. Hall, J.-M. Lehn, A. DeCian, J. Fischer, *Helv. Chim. Acta* **1991**, *74*, 1.
[38] M. Onishi, K. Isagawa, *Inorg. Chim. Acta* **1991**, *179*, 155.
[39] For a review of enantioselective transfer hydrogenations, see: G. Zassinovich, G. Mestroni, S. Gladiali, *Chem. Rev.* **1992**, *92*, 1051.
[40] (a) J. Tsuji, I. Minami, *Acc. Chem. Res.* **1987**, *20*, 140;
(b) B. M. Trost, T. R. Verhoeven, in *Comprehensive Organometallic Chemistry*, Vol. 8 (Eds.: G. Wilkinson, F. G. A. Stone, E. W. Abel), Pergamon, Oxford, **1982**, pp. 799-938; (c) B. M. Trost, *Acc. Chem. Res.* **1980**, *13*, 385; (d) C. G. Frost, J. Howarth, J. M. J. Williams, *Tetrahedron: Asymmetry*, **1992**, *3*, 1089.
[41] Reviews: G. Consiglio, R. M. Waymouth, *Chem. Rev.* **1989**, *89*, 257 and ref. [40d].
[42] (a) B. M. Trost, D.J. Murphy, *Organometallics* **1985**, *4*, 1143;
(b) P. R. Auburn, P.B. Mackenzie, B. Bosnich, *J. Am. Chem. Soc.* **1985**, *107*, 2033; P. B. Mackenzie, J. Whelan, B. Bosnich, *ibid.* **1985**, *107*, 2046;
(c) T. Hayashi, A. Yamamoto, T. Hagihara, Y. Ito, *Tetrahedron Lett.* **1986**, *27*, 191; T. Hayashi, A. Yamamoto, Y. Ito, E. Nishioka, H. Miura, K. Yanagi, *J. Am. Chem. Soc.* **1989**, *111*, 6301; T. Hayashi, *Pure Appl. Chem.* **1988**, *60*, 7;

(d) Y. Okada, T. Minami, Y. Sasaki, Y. Umezu, M. Yamaguchi, *Tetrahedron Lett.* **1990**, *31*, 3905; Y. Okada, T. Minami, Y. Umezu, S. Nishikawa, R. Mori, Y. Nakayama, *Tetrahedron: Asymmetry* **1991**, *2*, 667;
(e) M. Yamaguchi, T. Shima, T. Yamagishi, M. Hida, *Tetrahedron Lett.* **1990**, *31*, 5049; *Tetrahedron: Asymmetry* **1991**, *2*, 663;
(f) M. Sawamura, H. Nagata, H. Sakamoto, Y. Ito, *J. Am. Chem. Soc.* **1992**, *114*, 2586;
(g) B. M. Trost, D. L. Van Vranken, *Angew. Chem.* **1992**, *104*, 194; *Angew. Chem. Int. Ed. Engl.* **1992**, *31*, 228; B. M. Trost, D. L. Van Vranken, C. Bingel, *J. Am. Chem. Soc.* **1992**, *114*, 9327; B. M. Trost, L. Li, S. D. Guile, *ibid.* **1992**, *114*, 8745.
[43] (a) A. Togni, *Tetrahedron: Asymmetry* **1991**, *2*, 683;
(b) ref. [25], [30], and [7c].
[44] G. Koch, A. Pfaltz, unpublished results.
[45] B. M. Trost, S. J. Brickner, *Am. Chem. Soc.* **1983**, *105*, 568 and ref. [42a].
[46] M. Zehnder, M. Neuburger (Institute of Inorganic Chemistry, University of Basel); P. von Matt, A. Pfaltz, unpublished results.
[47] M. Sawamura, Y. Ito, *Chem. Rev.* **1992**, *92*, 857.
[48] See, e.g., D. M. P. Mingos, in *Comprehensive Organometallic Chemistry* (Eds.: G. Wilkinson, F. G. A. Stone, E. W. Abel), Pergamon, Oxford, **1982**, Vol. 3, Section 19.4.3.
[49] P. von Matt, A. Pfaltz, unpublished results.
[50] P. von Matt, A. Pfaltz, *Angew. Chem.* **1993**, *105*, 614.
[51] Systematic name: 2-[2-(diphenylphosphino)phenyl]-4,5-dihydrooxazole.
[52] A. I. Meyers, E. D. Mihelich, *J. Org. Chem.* **1975**, *40*, 3158; H. W. Gschwend, A. Hamdan, *J. Org. Chem.* **1975**, *40*, 2008.
[53] For analogous allylic aminations, see ref. [42c].

New Aspects in Stereoselective Synthesis of Aminoalcohols and Amino Acids

J. MULZER

Institut für Organische Chemie der Freien Universität Berlin,
Takustrasse 3, D-14195 Berlin

The total synthesis of rare, i.e. non-proteinogenic amino acids has received much attention during the past few years. One main reason is the discovery that small and hence synthetically accessible peptides have immense physiological activities and may thus be suitable as drugs after appropriate derivatization.

One striking example is the luteinizing hormone releasing hormone (LHRH) (Fig.1), a decapeptide isolated by Schally from porcine hypothalamus in 1977 [1].

All ten amino acids in LHRH are proteinogenic and hence amenable to straightforward derivatization, e.g. substitution by the (R)-analogues or close derivatives. The mode of action of LHRH is shown in Fig. 2. LHRH is produced in the hypothalamus. The LHRH receptor, located in the hypophysis, triggers the production of gonadotropic hormones (LH and FSH) which stimulate the sexual organs (testicles or ovaries) to generate the sexual hormones properly. In consequence any interaction of the LHRH receptor with LHRH agonists or antagonists (inhibitors) should have a deep influence on the activity of the sexual organs, and hence, in case of cancer, on the growth of malignant cells. This concept has led to a variety of synthetic LHRH antagonists which are all decapeptides with some of the natural proteinogenic amino acids exchanged for unnatural ones [2]. However, all these substitutes are quite similar to the natural amino

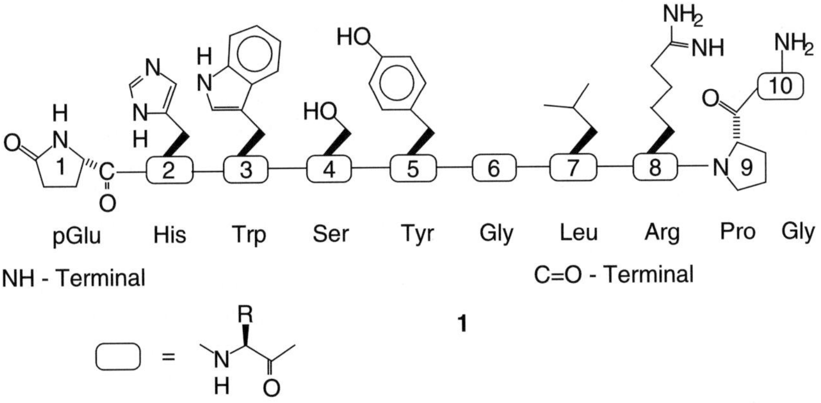

pGlu His Trp Ser Tyr Gly Leu Arg Pro Gly

NH - Terminal C=O - Terminal

1

Fig. 1. Luteinizing Hormone Releasing Hormone (LHRH)

Stereoselective Synthesis
Editors: Ottow, Schöllkopf, Schulz

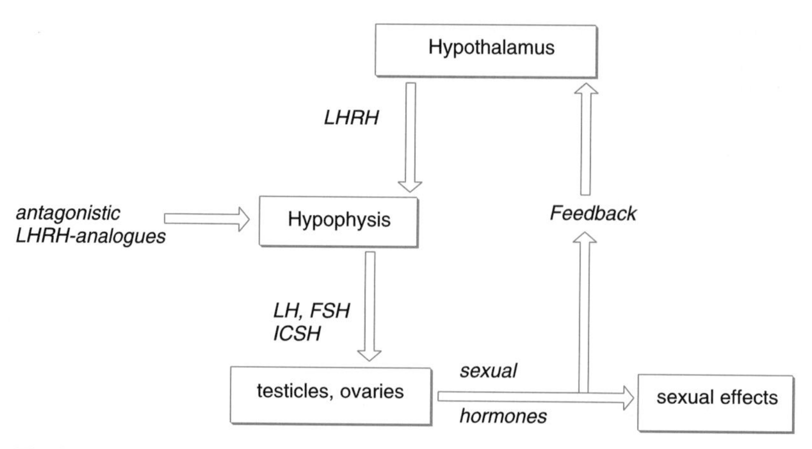

Fig. 2. Mode of Action of LHRH

acids, e.g. (R)-1-naphthalenyl alanine for (S)-tryptophane in the 3-position and the like. To take a significant step forward we decided to synthesize totally unnatural amino acids with defined geometry and functionality and to incorporate them into the LHRH structure. This lecture will deal with the synthesis of such amino acids and some of the corresponding amino alcohols.

There are four critical segments in the LHRH where amino acid substitution leads to high antagonistic effects: (a) the 1-3 segment, (b) the 6-position, (c) the 8-position and (d) the 9-10-position.

1. The 1-3 segment

In this segment aryl alanines have proven very effective in many cases. The meanwhile classical way to prepare aryl alanines is the diastereoselective alkylation of chiral glycine derivatives e.g. by the methods of Schöllkopf [3], Evans [4], Oppolzer [5], Seebach [6] and others. (Fig.4)

The method by Schöllkopf (Fig. 5) illustrates the advantages and disadvantages (Fig. 6) of these procedures.

The chiral glycinate anion **2** is alkylated with a potentially carcinogenic or lachrymatory benzylic halide to give the adduct **3**, usually as a diasteromeric mixture which has to be separated by high resolution chromatography. **3** has, then, to be cleaved into the auxiliary (here valine methyl ester (**4**)) and the desired amino ester 2-OMe-1-NAL (**5**), whose ee-value is 92 %. The pro and cons of this method are summarized in Figure 6.

The synthesis is short (6 steps including the preparation of **2**) but has a number of drawbacks, namely the necessity that the auxiliary has to be recovered and recycled, that problematic alkylating agents have to be employed, that the ee-values are relatively low (90-95 % typically), that very low reaction temperatures have to be applied (-100°C) and "hard" metallation conditions (butyllithium, strict exclusion of moisture and oxygen) are required.

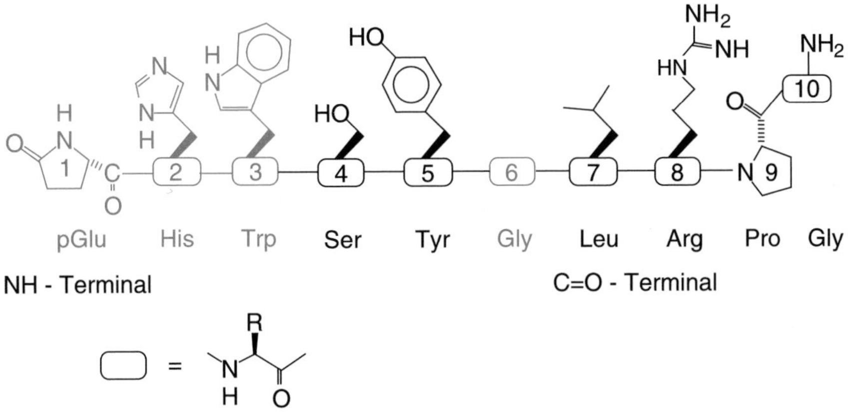

pGlu His Trp Ser Tyr Gly Leu Arg Pro Gly

NH - Terminal C=O - Terminal

Fig. 3. LHRH-Antagonists. Decapeptides with Unnatural Amino Acids: 4 Critical Segments

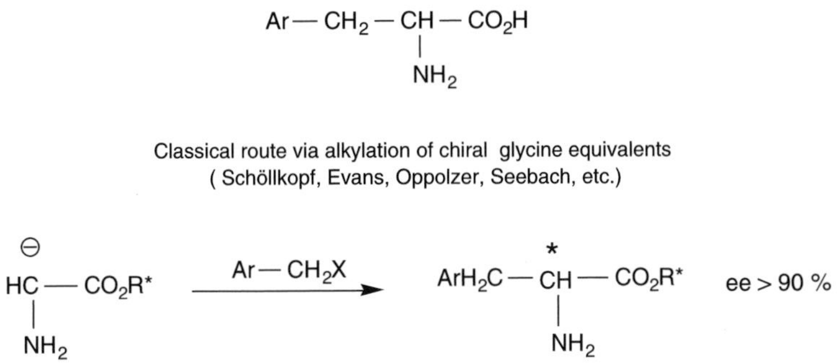

$$Ar-CH_2-CH(NH_2)-CO_2H$$

Classical route via alkylation of chiral glycine equivalents
(Schöllkopf, Evans, Oppolzer, Seebach, etc.)

$$HC^{\ominus}(NH_2)-CO_2R^* \xrightarrow{Ar-CH_2X} ArH_2C-\overset{*}{C}H(NH_2)-CO_2R^* \quad ee > 90\ \%$$

Fig. 4. Synthesis of Arylalanines for Pos. 1-3

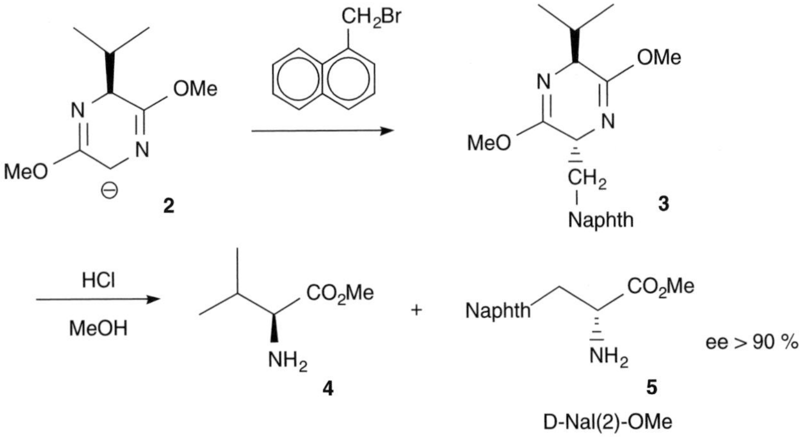

ee > 90 %

D-Nal(2)-OMe

Fig. 5. Aryl Alanine Synthesis According to Schöllkopf

Pro: short sequence

Cons:
- ° synthesis, removal and recovery of the auxiliary
- ° carcinogenic alkylating agents
- ° ee-values relatively low
- ° low temperatures
- ° "hard" metallation conditions

Fig. 6. Pro and Cons

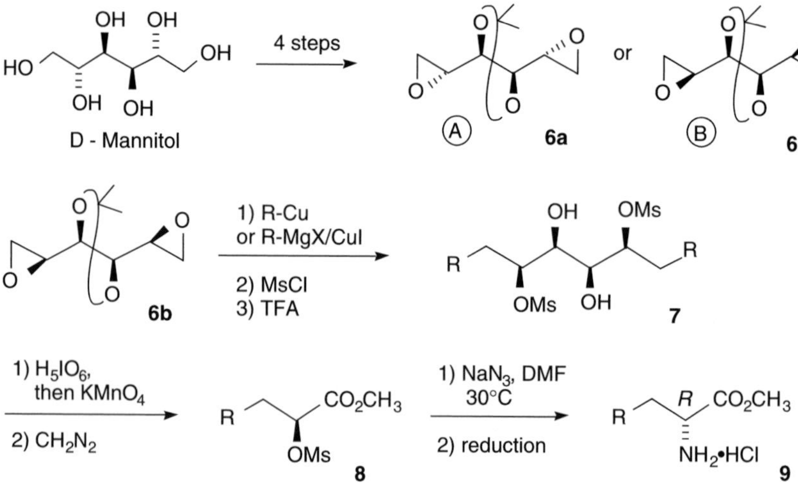

Fig. 7. Ex-Chiral-Pool-Synthesis of Aryl Alanines

Pros:
- ° no superbases required
- ° reaction temperature between 0°C and 30°C
- ° ee-values > 99.5%
- ° aryl halides easily accessible and harmless
- ° D- or L-amino acids accessible without changes of the reaction scheme
- ° total yields 20 - 53%!
- ° generally no chromatography required

Con:
- ° long sequence (11 steps)

Fig. 8. Pros and Con

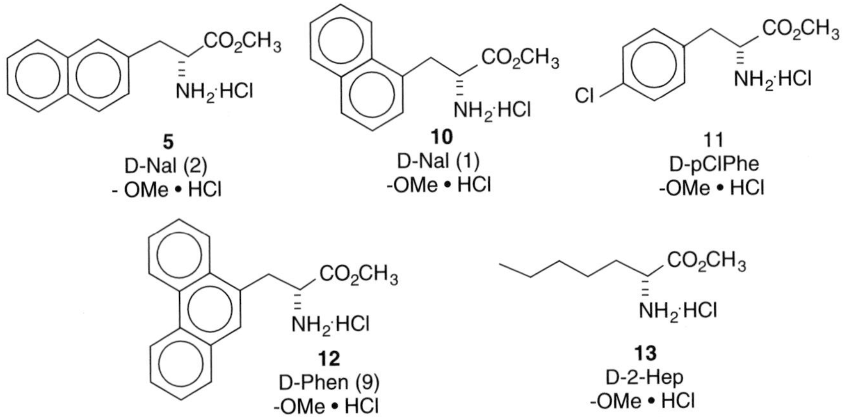
CO2CH3
NH2·HCl
5
D-Nal (2)
- OMe • HCl
CO2CH3
NH2·HCl
10
D-Nal (1)
-OMe • HCl
CO2CH3
Cl
NH2·HCl
11
D-pClPhe
-OMe • HCl
CO2CH3
NH2·HCl
12
D-Phen (9)
-OMe • HCl
CO2CH3
NH2·HCl
13
D-2-Hep
-OMe • HCl

Fig. 9. Some Examples for the Method Shown in Fig. 7

Our method (Fig.7) starts from the "chiral carbon pool" and uses the bis-epoxides **6a** and **b**, respectively, both available from D-mannitol in 4 steps [7]. **6b**, for instance, is opened regioselectively with organocuprate reagents at the terminal position to give, after bismesylation and removal of the acetonide, tetrol **7**. Oxidative cleavage and methylation furnishes the ester **8** without racemization which is then treated with sodium azide and reduced to give optically pure (ee > 99.5%) amino ester **9**.

The pros and con are listed in Fig. 8 and show, that the main drawback, i.e. the considerable number of synthetic steps is counterbalanced by considerable advantages. For instance no chromatographic separations are needed, the aryl halides are readily available and harmless and high ee-values are obtained with unfailing certainty.

Fig. 9 shows some examples, the list of which might be expanded almost arbitrarily. The only limitation lies in the reactivity of the aryl cuprates, which means that pyridyl or thienyl residues cannot be introduced yet. On the other hand the flexibility of the bisepoxide method is demonstrated by the synthesis of 2,3-diamino-propionic acid (β-amino-alanine) which is a constituent of antibiotic cyclo-peptides like tuberactinomycin A or, in form of the N-methyl derivative L-BMAA, occurs in the seeds of the local sago palm and had notoriously caused the "Guam disease" [8], with a certain kind of Alzheimer and Parkinson symptoms, until the American conquest of the island in World War II led to a change in nutrition (Fig. 10).

Although there is a number of syntheses of **14** [9] the main problem lies in the differentiability of the two amino functions, preferably in form of β-BOC and α-Fmoc-protection as shown in formula **16**. Our synthesis (Fig. 11) proceeds via the β-azido-α-mesyloxycarboxylic acid **17**, readily available from **6** by azide opening. Methylation and reduction of the azide function, followed by BOC protection, furnishes **19**, which after a second S_N2 displacement reaction with azide gives **20**. By ester hydrolysis, azide reduction and Fmoc-protection the desired bis-amino acid **16** is obtained in gram quantities.

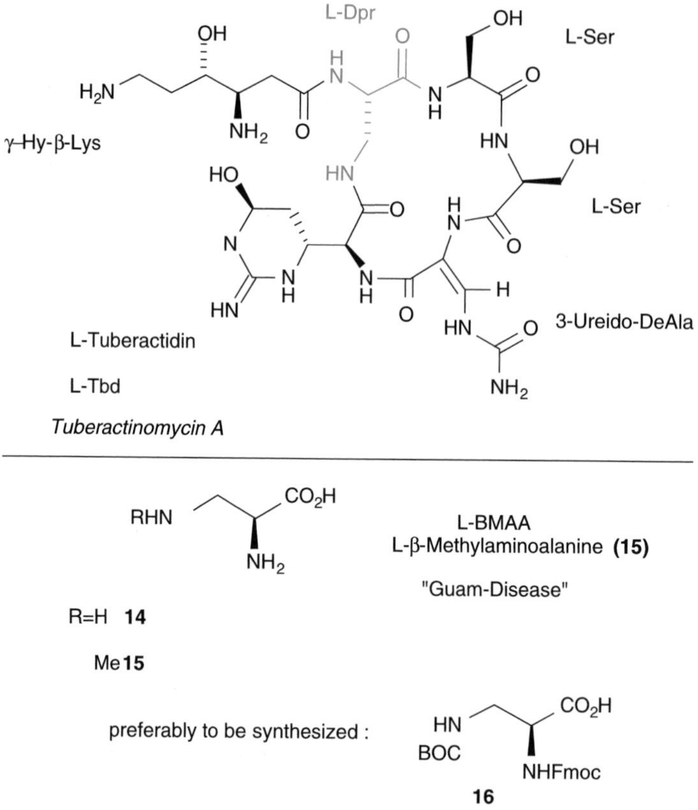

Fig. 10. β-Amino Alanines in Natural Products

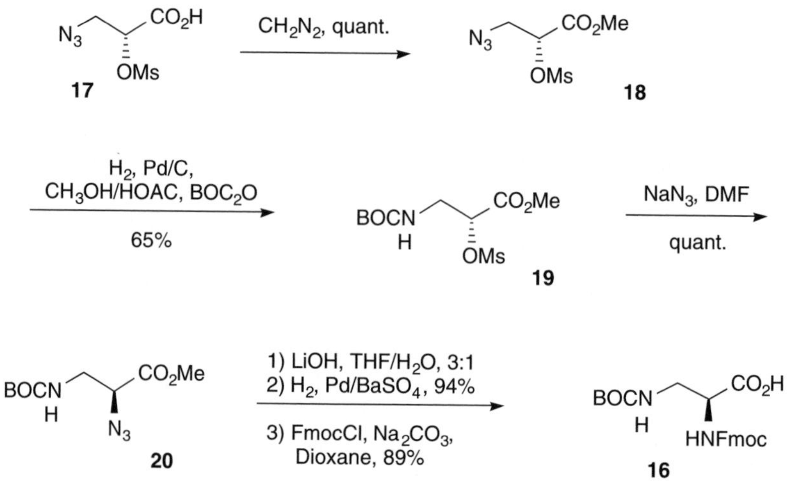

Fig. 11. Synthesis of **16**

2. The 6-Position

The amino acid in the 6-position exerts a decisive influence on the overall conformation of the LHRH antagonist.

Peptide conformations in general are determined by allyl-1,3-strain (Fig.12) [10] which means that only hydrogen is tolerated in the plane of the peptide bond. The R-groups alternately point up and down and the peptide strand adopts a pleated shape. In LHRH the 6-position has no R-group so that the conformation of this segment is not fixed. Hence **1** consists of two relatively rigid sections (1-5 and 7-10, respectively) mobile towards each other around the 6-glycinate as a central joint.

Clearly, the incorporation of a substituted amino acid into the 6-position could impose conformational rigidity on the molecule and hence, might give it a better fit into the corresponding receptor.

One could even combine this with the introduction of a basic or acidic end group in the 6-amino acid suitable for an interaction with some receptor function. This concept led to the synthesis of MorLys **21**, PipLys **22** and PyrLys **23** as basic amino acids and 2-amino suberinic acid **24** as an acidic one (Fig. 14).

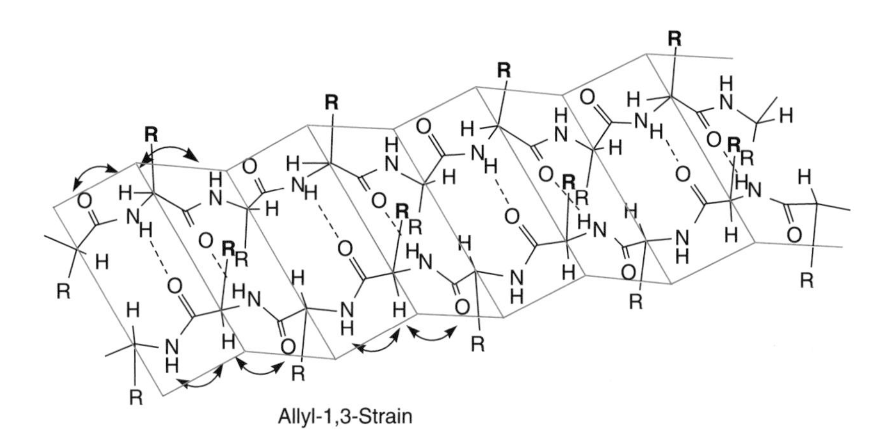

Fig. 12. 1,3-Allyl-Strain in Peptide Chains

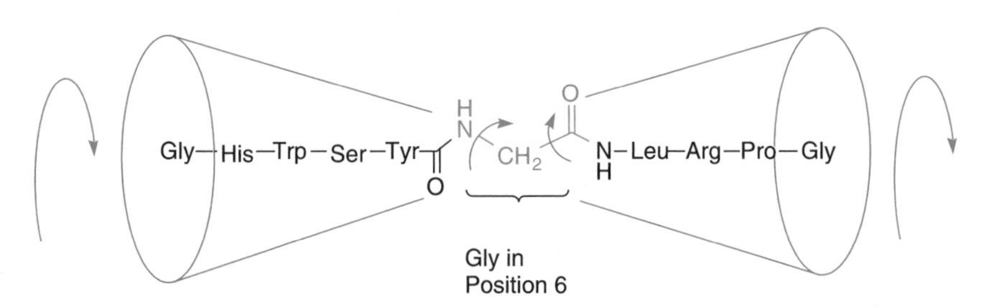

Fig. 13. Conformational Properties of LHRH

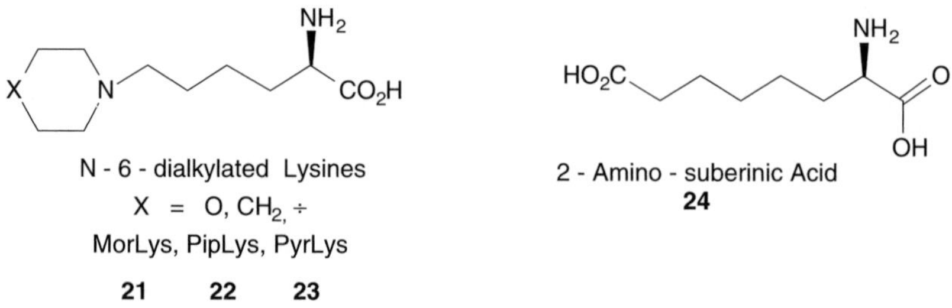

N - 6 - dialkylated Lysines
X = O, CH_2, ÷
MorLys, PipLys, PyrLys
21 **22** **23**

2 - Amino - suberinic Acid
24

Fig. 14. Amino Acids with Polar End Groups for Position 6

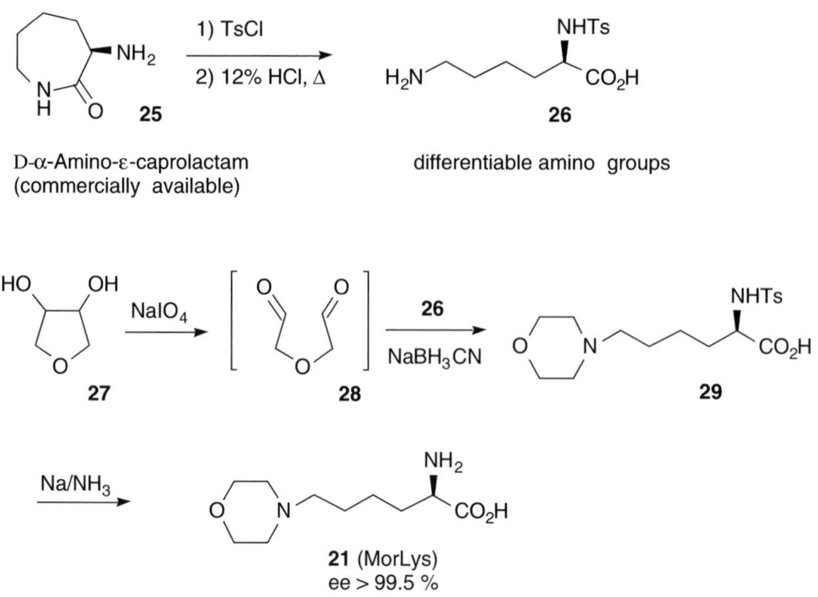

1) TsCl
2) 12% HCl, Δ

25
D-α-Amino-ε-caprolactam
(commercially available)

26
differentiable amino groups

$NaIO_4$
27 **28**
26
$NaBH_3CN$
29

Na/NH_3
21 (MorLys)
ee > 99.5 %

Fig. 15. Synthesis of MorLys

The central issue in the synthesis of **21**-**23** is the differentiation of the amino functions in lysine. This was achieved by starting from commercially available (R)- or (S)- **25** which can be tosylated and hydrolysed to give **26** without racemization. To introduce the heterocycle reductive amination was applied by generating the dialdehyde **28** from **27** and adding the reducing agent in situ along with **26**. In this way **29**, and after reductive removal of the tosyl group, MorLys **21** was obtained in 50-60 % overall yield (Fig. 15). Similarly, **22** and **23** were prepared.

The synthesis of **24** (Fig. 16) started with the Grignard addition to glyceraldehyde **30** which furnished **31** with good anti-selectivity. Mitsunobu reaction, diastereomer separation by crystallization and acetonide hydrolysis gave **32**, which was converted into **33** by oxidation and into **24** after hydrazinolysis.

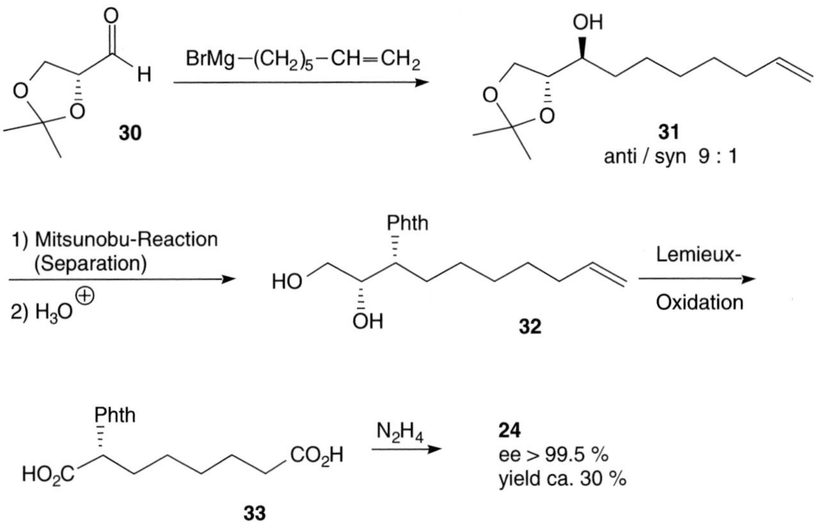

30 $\xrightarrow{BrMg-(CH_2)_5-CH=CH_2}$ 31 (anti / syn 9 : 1)

1) Mitsunobu-Reaction (Separation)
2) $H_3O^{\oplus}$
→ 32 (HO, Phth, OH) → Lemieux-Oxidation →

33 (HO_2C, Phth, CO_2H) $\xrightarrow{N_2H_4}$ 24
ee > 99.5 %
yield ca. 30 %

Fig. 16. 2-Aminosuberinic Acid

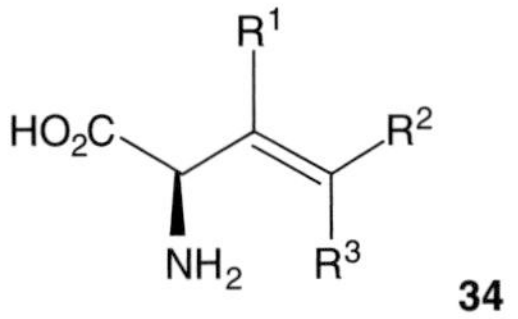

34 (HO_2C, NH_2, R^1, R^2, R^3)

Aim: Incorporation in the LHRH-6-position. Subsequent modification of the C=C-double bond or in Position 1-3 as a substitute for NAL etc.

also: Application as Suicide Inhibitor (Plant Protection)

Fig. 17. Synthesis of β,γ-Unsaturated α-Amino Acids

A different possibility of 6-functionalization is the introduction of unsaturated amino acids like **34** (Fig. 17) whose olefinic function could be functionalized after incorporation into the decapeptide.

For the synthesis of **34** aldehyde **30** was allylated to give **35**. Base catalyzed double bond migration led to **36** and after debenzylation to **37**. Now the nitrogen had to be introduced via Mitsunobu reaction to give **38** which could then be transformed into the desired amino acid **39**.

The crucial point is the potential allylic transposition in the Mitsunobu reaction, i.e. the regio- and stereochemical outcome of the attack of **41** on the activated allylic alcohol **40** (Fig. 19).

Our results are summarized in Fig. 20. Only if relatively stable carbenium ions (sec.-sec.) like **43** or **47** can be formed, allylic transposition is observed. The α/γ-ratio is then controlled by the relative degree of steric hindrance at both reactive sites.

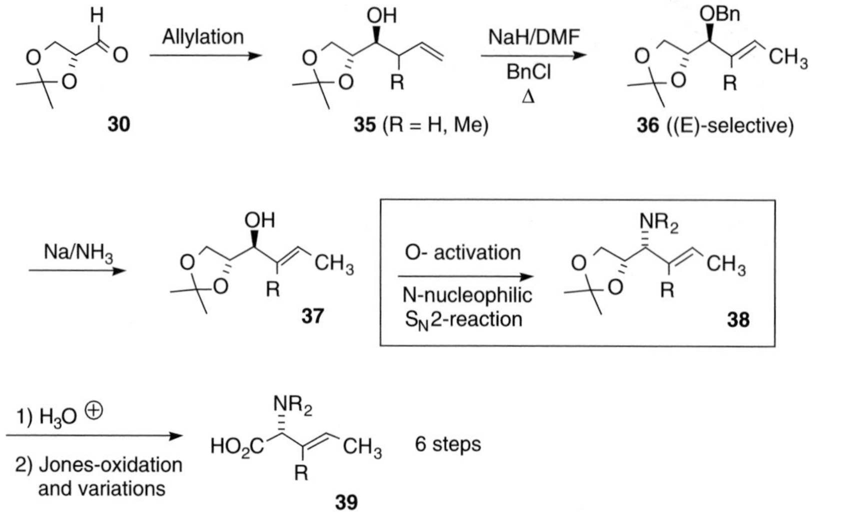

Fig. 18. S_N2-Introduction of Nitrogen

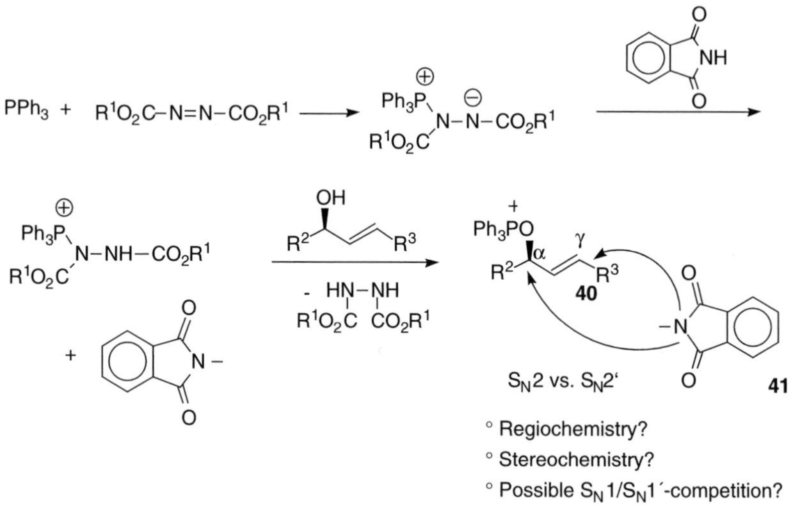

Fig. 19. S_N2/S_N2'-Competition in the Mitsunobu-Reaction

In the case of **44**, the corresponding carbenium ion **45** is too unstable to be formed, and S_N2-product only is obtained in this case. The γ-attack occurs mainly anti to the α-OH.

However, the relative configurations of the α-carbinol and the acetonide-O also play a major role. Thus, in **48**, with an anti-arrangement of this 2,3-diol moiety, the (R)/(S)-ratio at the γ-position of **49** is only 75:25, whereas in **50** (syn-arrangement of the 2,3-diol) it is 4:96 ! (Fig. 21). An explanation is given in Fig. 22.

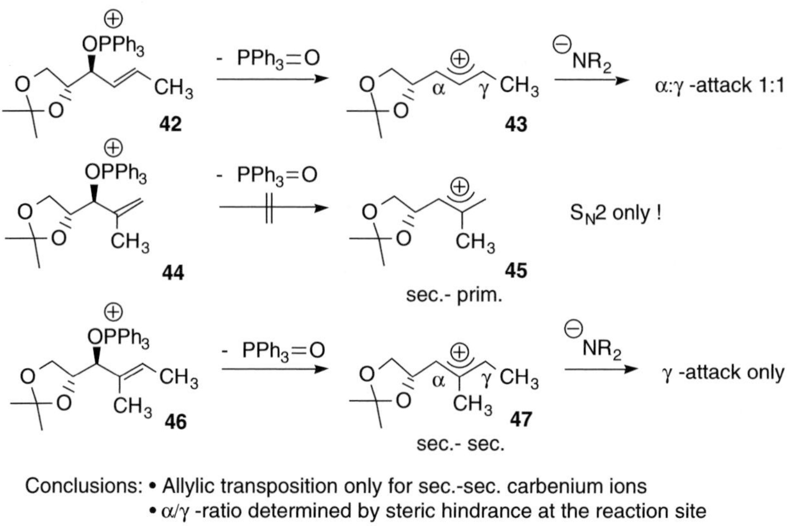

Conclusions: • Allylic transposition only for sec.-sec. carbenium ions
• α/γ -ratio determined by steric hindrance at the reaction site

Fig. 20. Allylic Rearrangement Follows Carbeniumion Stabilization: Hence S_N1'-Mechanism

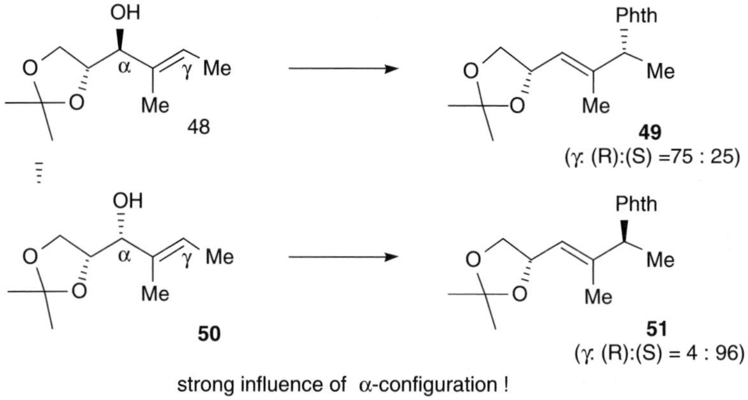

strong influence of α-configuration !

Fig. 21. Stereochemistry of the γ-Attack

The Mitsunobu reaction proceeds via the activated oxyphosphonium derivatives **48a** and **50a**, respectively. From the (E)-configuration of **49** and **51** we know that the reactive conformations of **48a/50a** must be exo as shown. From the configuration of the products we conclude that the nucleophile attacks anti to the leaving group. For **48a** this means that the nucleophile has to cope with the dipole repulsion of the acetonide oxygens. **50a**, by contrast, is devoid of such a repulsion. That the acetonide oxygens indeed exert such a repulsive effect is also demonstrated by the regiochemistry of the reaction (Fig. 23).

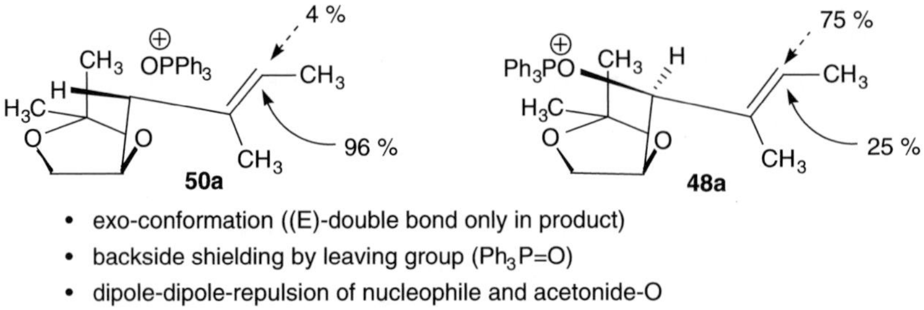
4 %
CH3 OPPh3
H
H3C
O
O
CH3
CH3
96 %
50a
Ph3PO
CH3
H
H3C
O
O
75 %
CH3
CH3
25 %
48a
• exo-conformation ((E)-double bond only in product)
• backside shielding by leaving group (Ph3P=O)
• dipole-dipole-repulsion of nucleophile and acetonide-O

Fig. 22. Stereochemistry of the S_N1'-Reaction

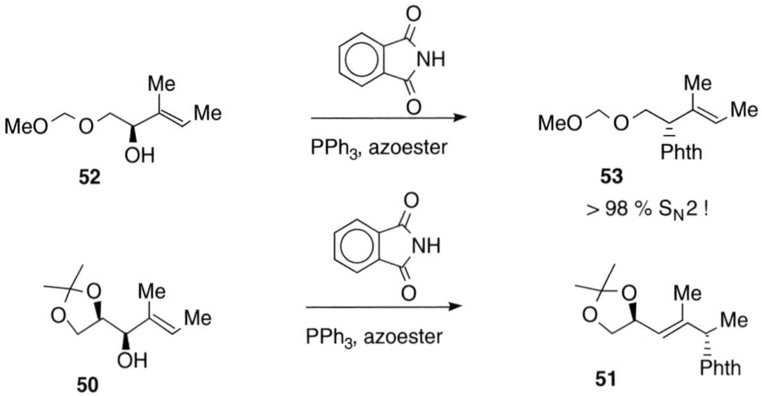
MeO
O
Me
Me
OH
52
NH
PPh3, azoester
Phth
53
> 98 % SN2 !
50
51

Fig. 23. Regio-Influence of the Acetonide

Whereas in **50** complete allylic transposition is observed, there is no such effect in **52,** which undergoes normal S_N2-displacement to give **53**.

The preparative consequences of these findings are shown in Fig. 24: amino acids **54** and **55** are easily accessible on the route described. **56** though not available from acetonide **50** can be prepared from the MOM-protected derivative **52.**

The α,β-unsaturated γ-amino acids **57** generally cannot be obtained from the S_N1'-route due to the lacking stereocontrol of the γ-attack. They are, however, easily prepared on a different route (Fig. 25) starting from lactic ester **58**. In the Mitsunobu substitution **61** → **62** S_N2-displacement is observed, exclusively because carbenium ions are unstable in α-position to an ester function.

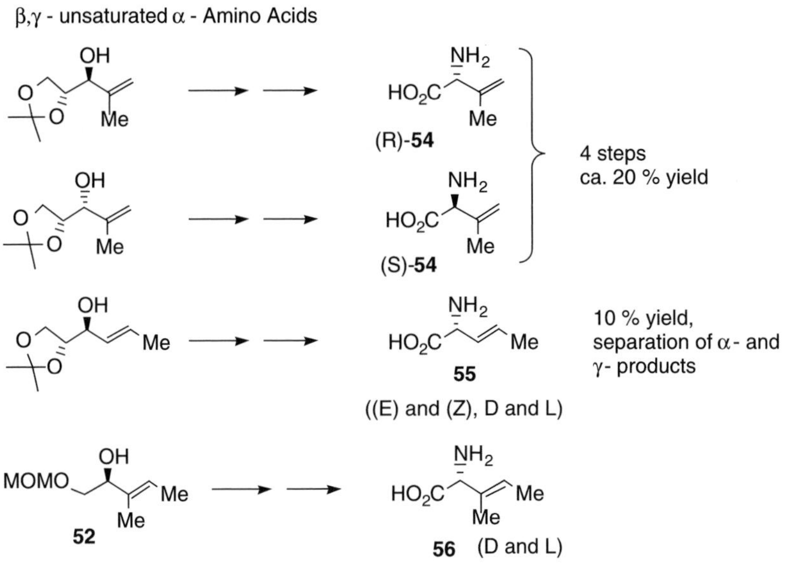

β,γ - unsaturated α - Amino Acids

OH → → NH_2 HO_2C Me (R)-**54**

OH → → NH_2 HO_2C Me (S)-**54**

4 steps
ca. 20 % yield

OH Me → → NH_2 HO_2C Me **55** ((E) and (Z), D and L)

10 % yield,
separation of α- and γ- products

MOMO OH Me Me **52** → → NH_2 HO_2C Me Me **56** (D and L)

Fig. 24. Preparative Consequences

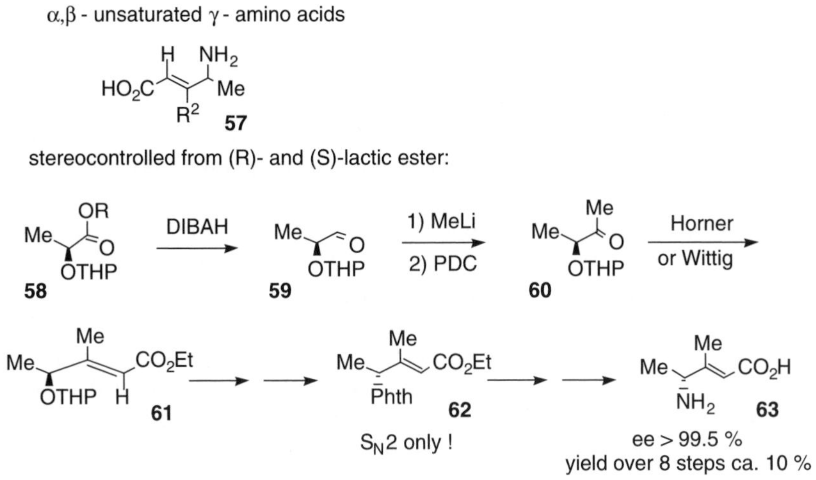

α,β - unsaturated γ - amino acids

H NH_2 HO_2C Me R^2 **57**

stereocontrolled from (R)- and (S)-lactic ester:

Me OR O OTHP **58** —DIBAH→ Me O OTHP **59** —1) MeLi 2) PDC→ Me Me O OTHP **60** —Horner or Wittig→

Me Me CO_2Et OTHP H **61** → → Me Me CO_2Et Phth **62** S_N2 only ! → → Me Me CO_2H NH_2 **63**

ee > 99.5 %
yield over 8 steps ca. 10 %

Fig. 25. Synthesis of α, β-unsaturated γ-amino acids via Mitsunobu pathway

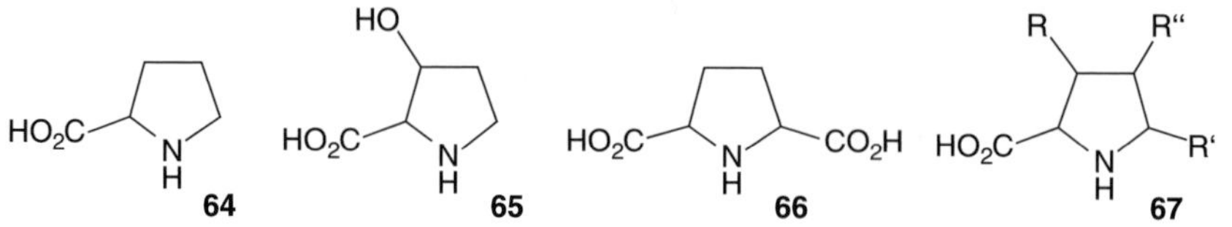

Fig. 26. Synthesis of Proline-Derivatives and Prolinols

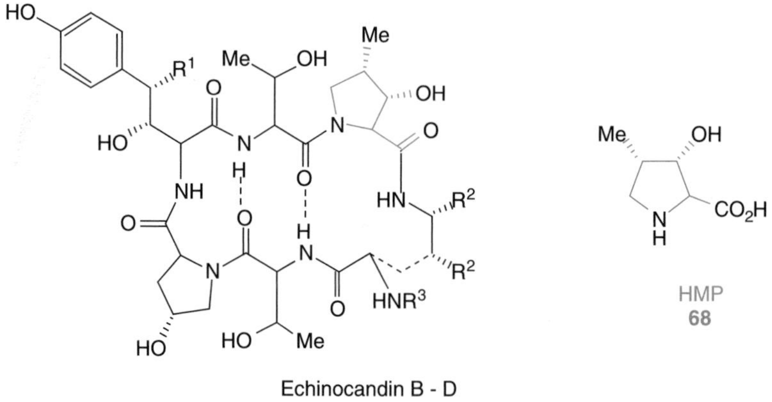

Echinocandin B - D

Fig. 27. Example: 3-Hydroxy-4-methylproline (HMP)

3. Position 9: Prolines

For the modification of the 9-position a variety of proline derivatives (e.g. **65-67**, Fig. 26) was envisaged, e.g. HMP (3-hydroxy-4-methylproline **68**), which is a constituent of the antibiotics echinocandin B, C and D (Fig. 27) [11].

For the synthesis of **68** the Staudinger cyclization shown in Fig. 28 was developed as a new method. Azido-epoxides **69** and **71** were prepared from D-mannitol in 8 steps and treated with triphenylphosphane to give, under elimination of nitrogen and triphenylphosphane oxide, the bicyclic amines **70** and **72**. The mechanism of the reaction (Fig. 29) shows that primarily the phosphine imine **73** is formed, which opens the epoxide to give **74/77**. N-O-migration of the phosphonium moiety generates **75/78**, which cyclizes under elimination of phosphine oxide to give **76** in a regioconvergent manner. To convert **70** into **68**, a regioselective opening of the aziridine ring at the exocyclic position had to be achieved. We discovered that the desired regiocontrol can be exerted by the choice of the anhydride.

Thus, benzoic acid anhydride converts **70** into **79**, whereas BOC-anhydride generates **80** (Fig. 30). From **79**, HMP was prepared via straightforward functional group manipulations (Fig. 31) [11].

To our surprise also unactivated azido-1,2-diols like **83-85** (Fig. 32) undergo Staudinger cyclizations in low yields. Better results were obtained from the cy-

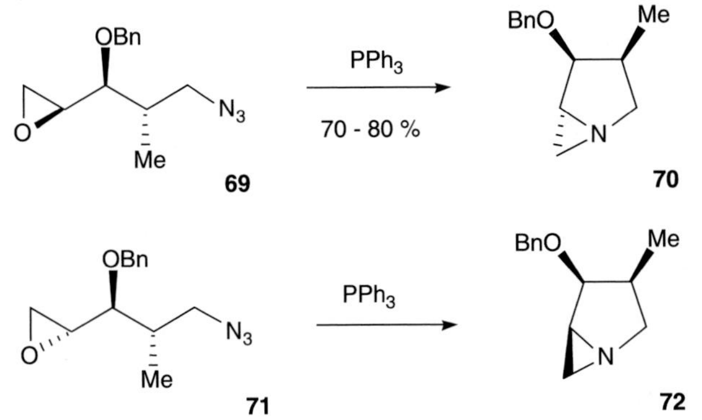

Fig. 28. Staudinger Cyclization

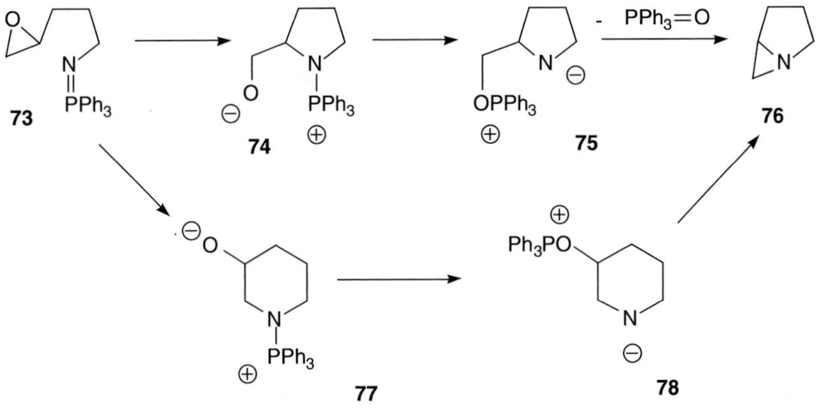

Fig. 29. Mechanism of the Staudinger-Cyclization

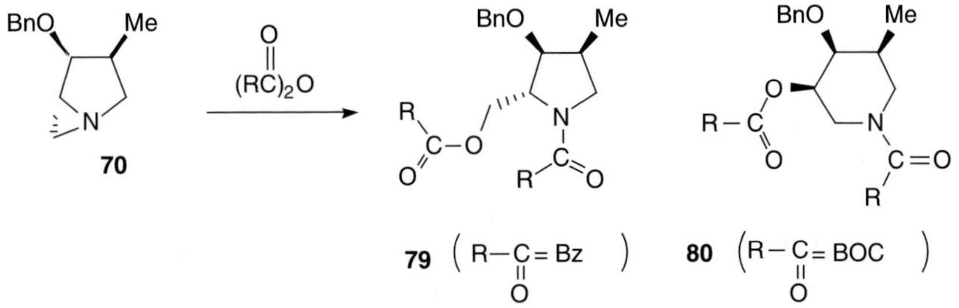

Fig. 30. Regioselectivity of the Ring Opening

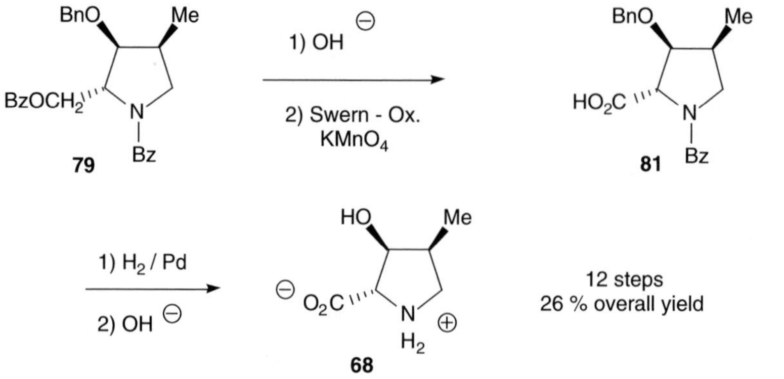

Fig. 31. Synthesis of HMP (**68**)

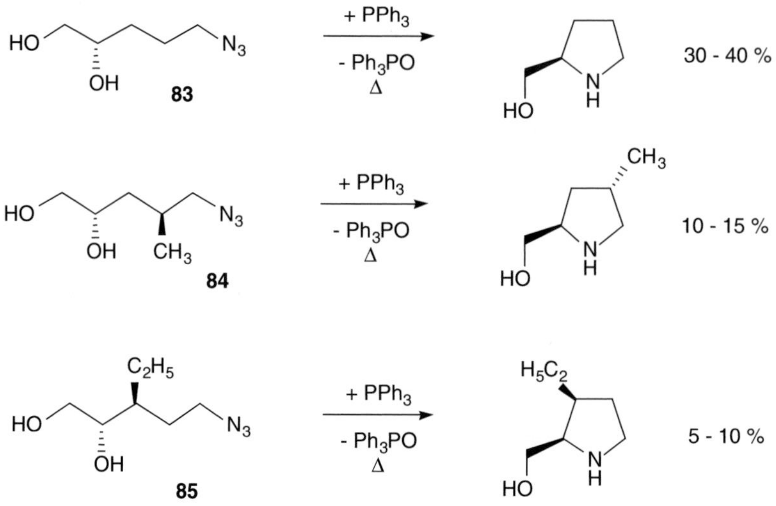

Fig. 32. Direct Staudinger-Cyclization of 1,2-Diols

clization of azido-aldehydes **86** which gave cyclic imines **88** via phosphine imine **87** for a wide variety of substituents (Fig. 33)

For instance, **89** was converted into **90** in 96 % yield, which could be transformed into **91-93** by nucleophilic additions to the C=N-bond. The nitrile function of **92** can be hydrolyzed to the amino acid, and the olefinic moiety in **91-93** may be submitted to a manifold of addition reactions (Fig. 34). After N-protection and O-deprotection prolinols **91-93** can be oxidized to prolines.

Acetyl azides like **94** can be cyclized to dihydrooxazoles **95** in high yield (Fig. 35). Furthermore, the protocol can be extended to six-membered rings as shown in Fig. 36 for the synthesis of pipecolic acid **100** from azido-aldehyde **96**.

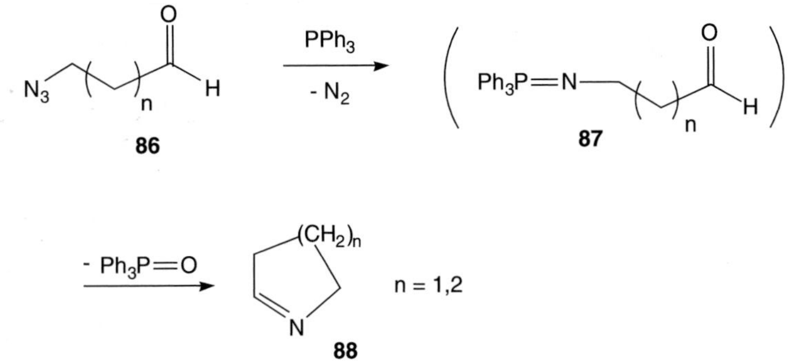

N_3 (n) H, O — **86**

PPh_3, $- N_2$ →

($Ph_3P{=}N$ (n) H, O — **87**)

$- Ph_3P{=}O$ →

$(CH_2)_n$, N — **88** n = 1,2

Fig. 33. Aza-Wittig-Staudinger-Cyclization

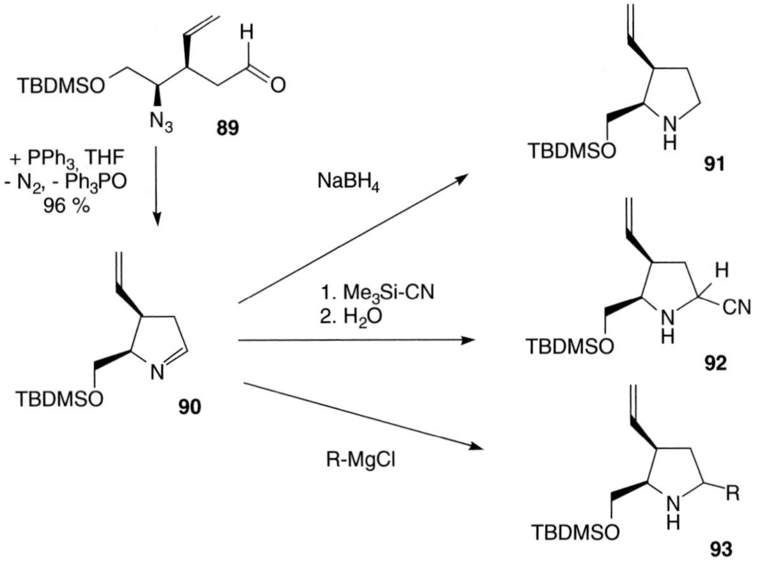

TBDMSO, N_3, H, O — **89**

$+ PPh_3$, THF
$- N_2$, $- Ph_3PO$
96 %

TBDMSO, N — **90**

$NaBH_4$ → TBDMSO, N H — **91**

1. Me_3Si-CN
2. H_2O → TBDMSO, N H, H, CN — **92**

R-MgCl → TBDMSO, N H, R — **93**

Fig. 34. Preparation of Prolinols **91-93**

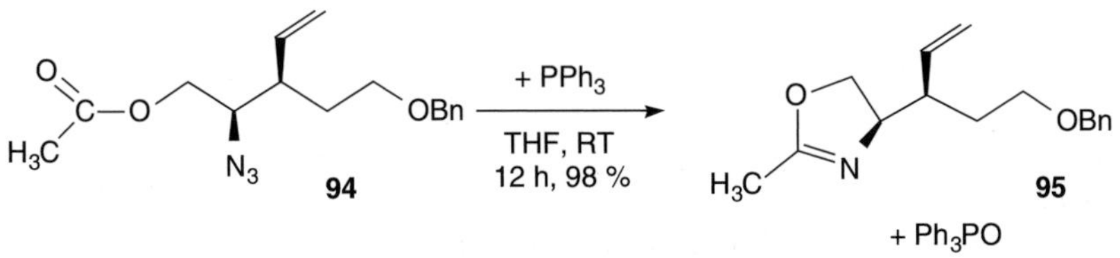

O, C—O, H_3C, N_3, OBn — **94**

$+ PPh_3$
THF, RT
12 h, 98 % →

O, N, H_3C, OBn — **95**

$+ Ph_3PO$

Fig. 35. 4,5-Dihydrooxazole-Synthesis

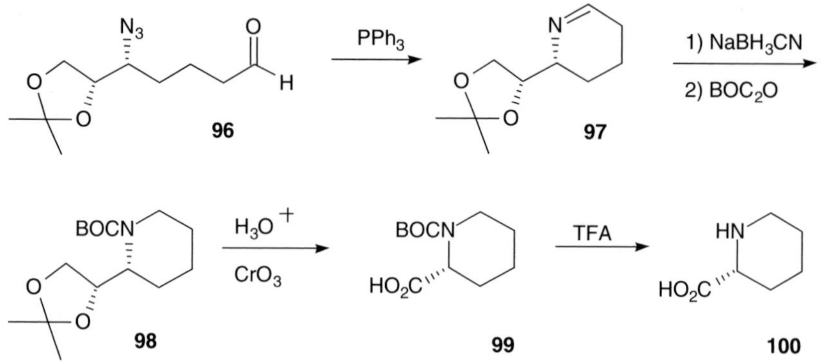

Fig. 36. Synthesis of Pipecolic Acid

4. Position 6, 8 or others: BAZ

The benzodiazepine moiety has proven its high pharmacological activity in many respects as the examples in Fig. 37 show [12]. Therefore it seemed attractive to incorporate the benzodiazepine structure also in a peptide chain so that the heterocycle has a defined orientation with respect to the peptide backbone and may interact with a suitable receptor protein. In principle the distance of the benzodiazepine from the peptide strand may be regulated by the number of the carbon atoms between the nitrogen atoms in the peptide bond and the endocyclic N-4. We chose the 1,3-distance and took BAZ (*b*enzodi*az*epino alanine **101**) as

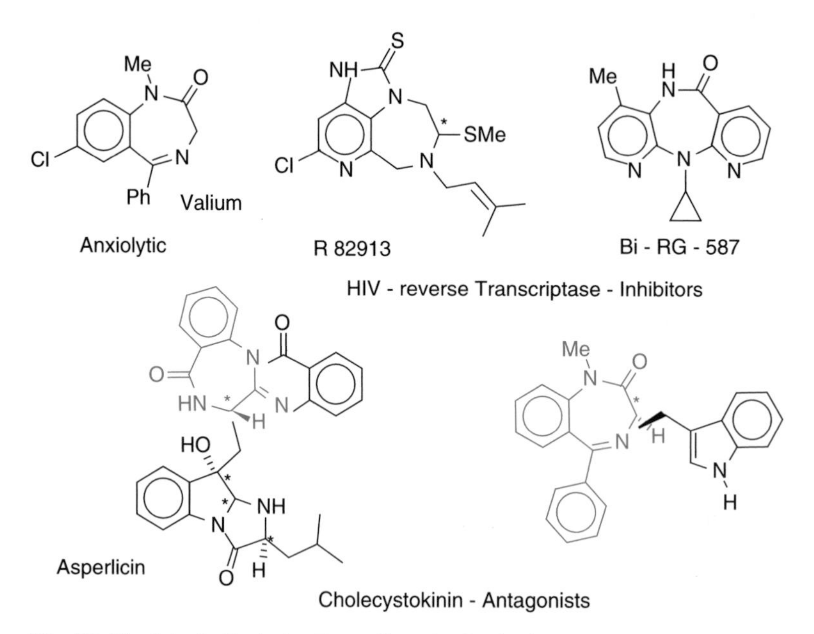

Fig. 37. Physiologically Active Benzodiazepine Derivatives

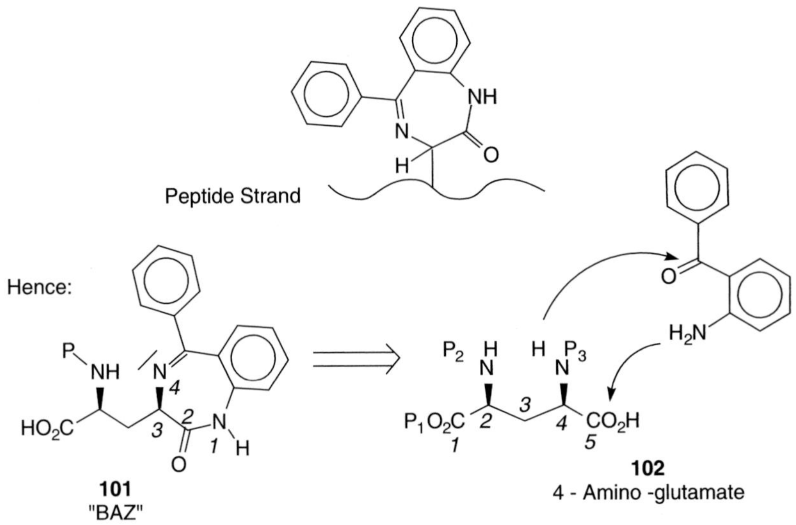

Peptide Strand

Hence:

101 "BAZ"

102 4 - Amino -glutamate

Fig. 38. Peptide Bound Benzodiazepines

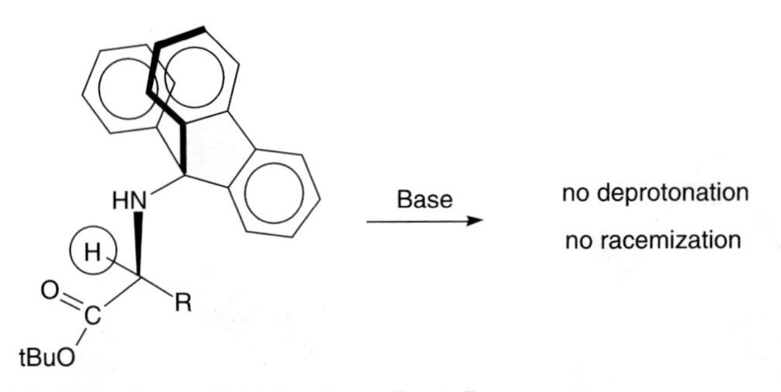

Fig. 39. α-Proton Shielding According to Rapoport

our target molecule. The two stereogenic centers may have syn or anti configuration and we intended to prepare both. Thus 4-amino glutamate **102** had to be synthesized as syn and anti diastereomer, and, furthermore, both amino acid moieties had to be differentiable so that one could be incorporated into the benzodiazepine and the other one in the peptide. For the syn arrangement this differentiation was also vital to avoid the meso-situation and hence the loss of optical activity (Fig.38). If, for instance, the 1,2-amino acid moiety is protected and the 4,5-amino acid moiety is unprotected, the benzodiazepine can be annelated to the 4,5-section by a Sternbach cyclization [13].

The differentiation was achieved by starting from L-glutamate and protecting the C-H-acidic 2-position against deprotonation by Rapoport's 9-phenylfluorenyl (9-Phf)-amino protective group [14]. If the adjacent ester is *tert*-butoxy, most superbases are too bulky to attack the corresponding hydrogen (Fig. 39).

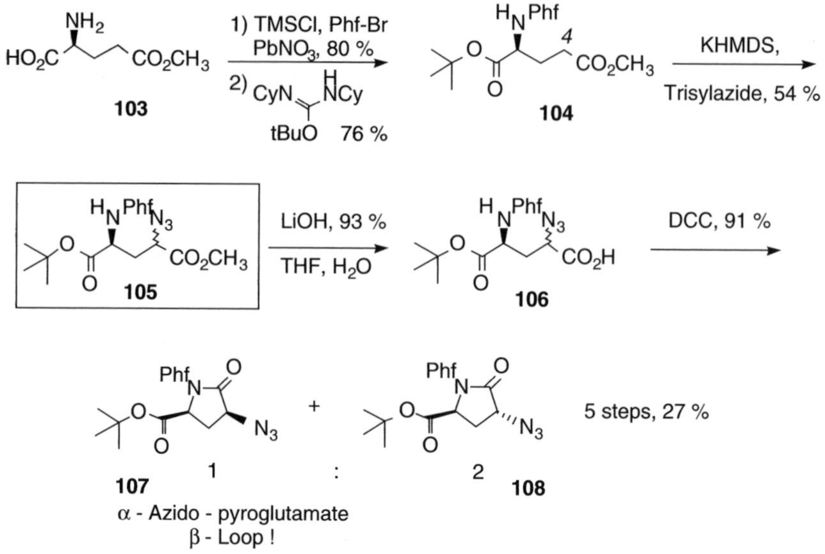

NH_2 HO_2C CO_2CH_3 **103**

1) TMSCl, Phf-Br $PbNO_3$, 80 %
2) CyN, NCy, tBuO 76 %

H_N Phf, 4, CO_2CH_3, O **104**

KHMDS, Trisylazide, 54 %

H_N Phf N_3 CO_2CH_3 O **105**

LiOH, 93 %
THF, H_2O

H N Phf N_3 CO_2H O **106**

DCC, 91 %

Phf N O N_3 O **107** 1 : 2 **108** Phf N O N_3 O

+

5 steps, 27 %

α - Azido - pyroglutamate
β - Loop !

Fig. 40. Synthesis of BAZ

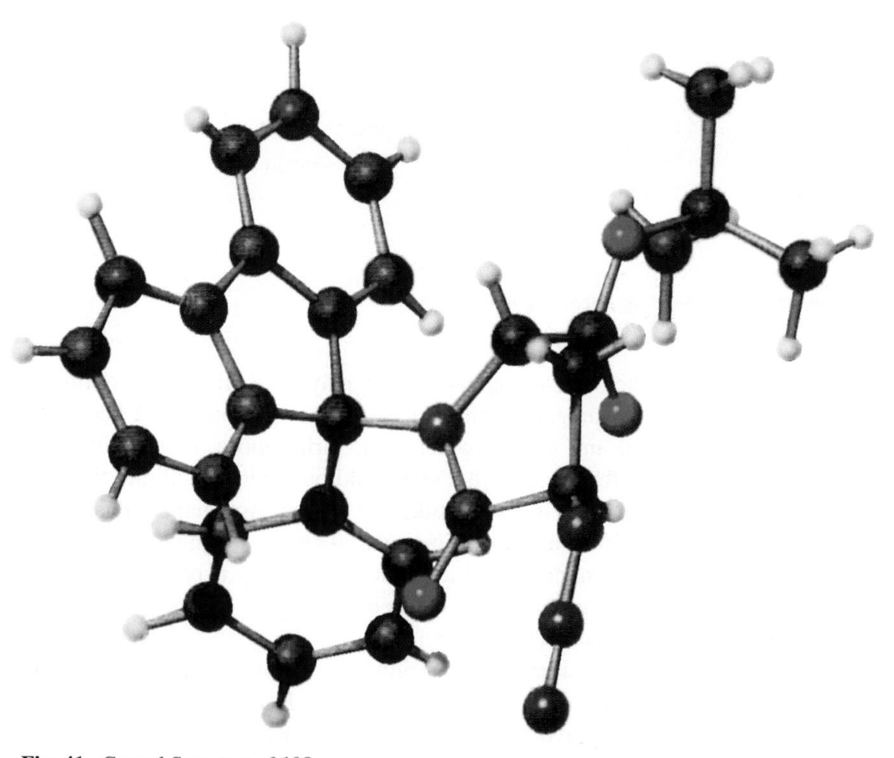

Fig. 41. Crystal Structure of **108**

Indeed, glutamate **104** (Fig. 40) can be selectively deprotonated at the 4-position and treated with Evans' trisylazide [4] to give **105** as a 2:1-diastereomeric mixture which is converted into the α-azido-pyroglutamates **107/108**.

These cyclic derivatives can be separated by chromatography. **108** is crystalline and submitted to an X-ray crystal structure analysis (Fig. 41).

One can see very clearly how the 9-Phf-group blocks the 2-H. Also, the azide function and the ester are rigidly fixed in an antiperiplanar conformation which can be incorporated in a peptide chain. Analogously the cis-isomer **107** could be used for generating a rigid β-turn in a peptide.

The synthesis of BAZ was continued by hydrogenating **105** and protecting the amino groups with BOC to give **109**. The differentiation of the ester groups is possible by a selective saponification of the methyl ester to acid **110** which is converted into the amide **111** (Fig. 42).

Deprotection of the amino groups followed by cyclization and Fmoc-protection of the exocyclic amino group delivers the desired BAZ derivative **114**, and quite analogously, **115** (Fig. 42).

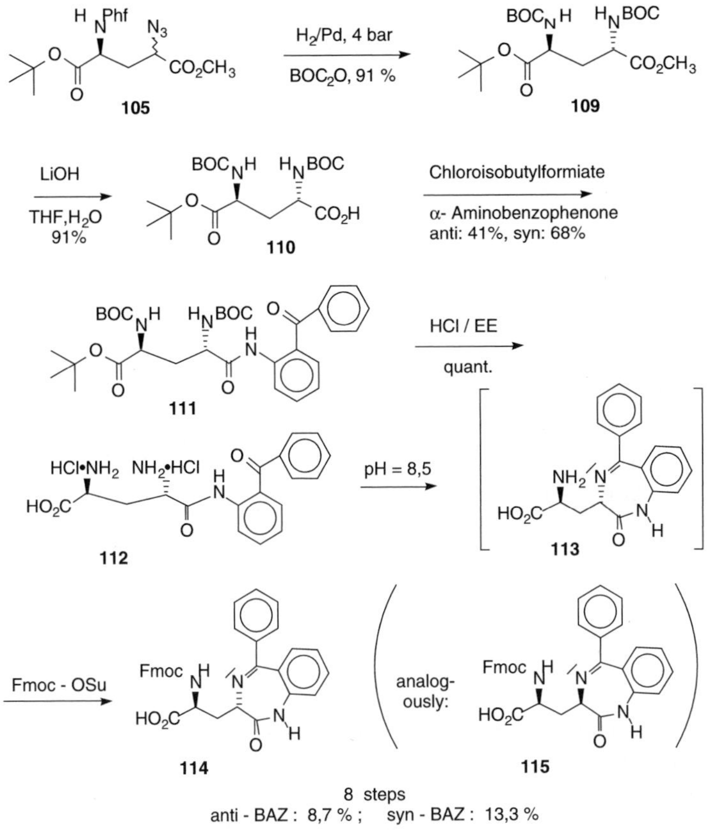

105 → H_2/Pd, 4 bar; BOC_2O, 91 % → 109

LiOH; THF,H_2O 91% → 110

Chloroisobutylformiate; α- Aminobenzophenone anti: 41%, syn: 68% → 111

HCl / EE; quant. → 112

pH = 8,5 → [113]

Fmoc - OSu → 114 (analogously: 115)

8 steps
anti - BAZ : 8,7 % ; syn - BAZ : 13,3 %

Fig. 42. Synthesis of BAZ

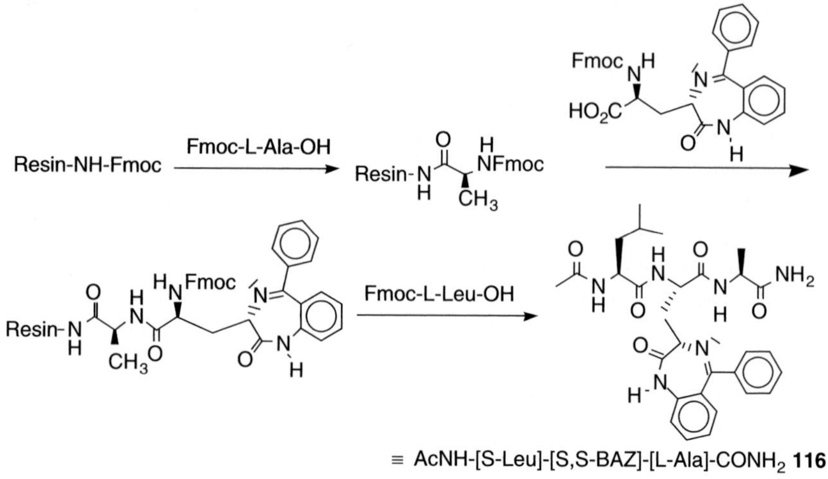

Resin-NH-Fmoc → (Fmoc-L-Ala-OH) → Resin-NH-CO-CH(CH_3)-NHFmoc → → (Fmoc-L-Leu-OH) →

≡ AcNH-[S-Leu]-[S,S-BAZ]-[L-Ala]-$CONH_2$ **116**

Fig. 43. Coupling of BAZ, Ala and Leu to Tripeptide **116**

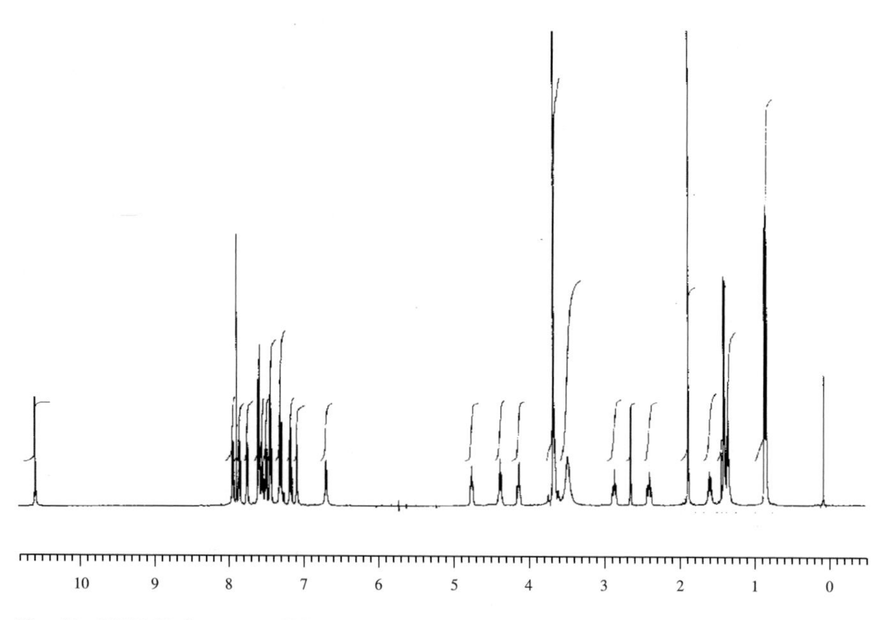

Fig. 44. ^{1}H-NMR Spectrum (500 MHz, $CDCl_3/CD_3OD$ 5:1) of **116**

To demonstrate that **114** is suitable for peptide coupling, it was incorporated in a tripeptide **116** by Dr. Lobbia (Schering AG, Berlin), via the Merrifield solid phase procedure (Fig. 43).

The 500 MHz-^{1}H-NMR spectrum of **116** (Fig. 44) shows the diasteromeric purity of the tripeptide, and hence the enantiomeric integrity of **114**. The surpris-

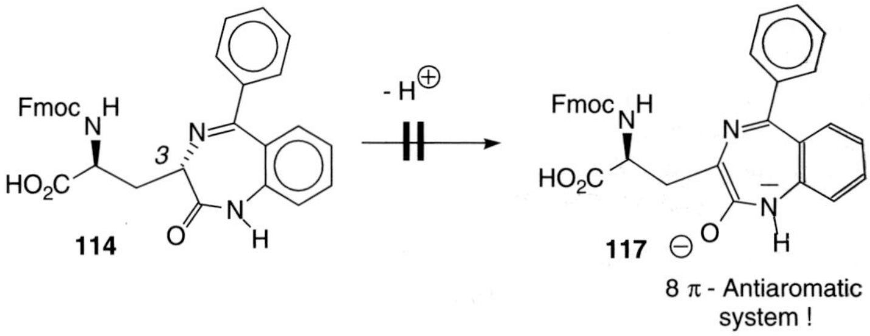

Fig. 45. Configurative Stability of BAZ

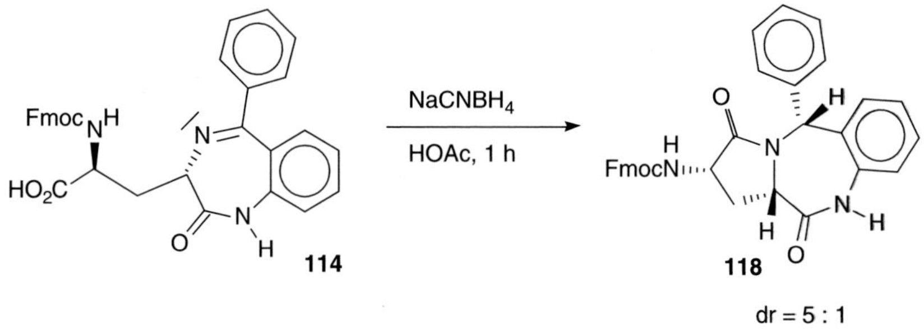

Fig. 46. Dihydro-BAZ-Lactam

ing configurational stability of the 3-position in the benzodiazepine may be interpreted by the fact that the corresponding anion **117** would form an 8-π-antiaromatic system (Fig. 45).

Due to this configurational stability **114** may also be hydrogenated at the C=N-bond with sodium cyanoborohydride (Fig. 46). The resulting amine spontaneously cyclizes to give lactam **118** with a diastereoselectivity of 5:1, which may be explained by the boat conformation of the heterocyclic ring facilitating the attack from the β-face.

5. Conclusion

As a conclusion a couple of guidelines and general statements have emerged from our amino acid work which are summarized in Fig. 47.

It is the hope of the author that these theses may be the basis of a controversial and fruitful discussion about the basic principles of asymmetric natural product synthesis. Finally, the author expresses his gratitude to his enthusiastic and capable coworkers (Fig. 48), whose experimental skill and diligence made these results possible.

- Syntheses with ee>99.5% only are acceptable
- there is no "allround" method covering all cases
- it is best to start from the "chiral carbon pool"
- the number of steps is not so important
- "simple" amino acids should be synthesized without changing the oxidation level
- "complex" amino acids are well accessible from polyols and subsequent oxidation to the carboxylic acid

Fig. 47. Statements Regarding the Synthesis of "Unnatural" Amino Acids

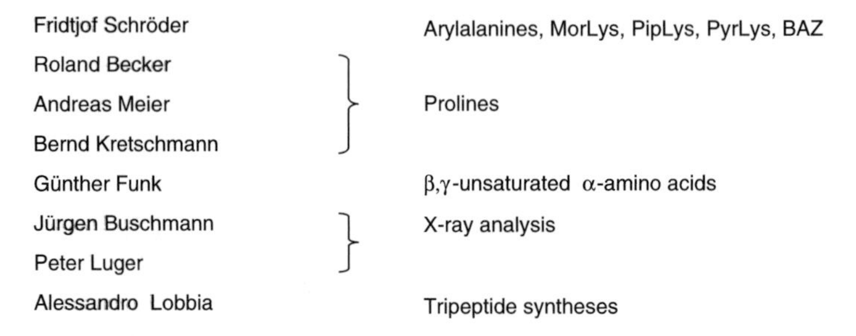

Fridtjof Schröder
Arylalanines, MorLys, PipLys, PyrLys, BAZ
Roland Becker
Andreas Meier
Bernd Kretschmann
Prolines
Günther Funk
β,γ-unsaturated α-amino acids
Jürgen Buschmann
Peter Luger
X-ray analysis
Alessandro Lobbia
Tripeptide syntheses

Fig. 48. Who Did the Experiments?

References

[1] H. Matsuo, Y. Baba, R.M. Nair, A. Arimura, A.V. Schally; *Biochem. Biophys. Res. Commun.* **1971** *43*, 1334

[2] a) A.S. Dutta, *Drugs of the future*, **1988** *13,* 761
b) R.C. Allen, *Annual Reports in med. Chem.* **1988** *23,* 210 c) A. Ljungqvist, D. Feng, C. Bowers, W.A. Hook, K. Folkers, *Tetrahedron* **1990** *46*, 3297

[3] U. Schöllkopf, *Top. Curr. Chem.* **1983** *109*, 65

[4] D.A. Evans, T.C. Britton, J.A. Ellman, R.L. Dorow, *J. Am. Chem. Soc.* **1990** *112*, 4011

[5] W. Oppolzer, R. Moretti, *Helv. Chim. Acta* **1986** *69*, 1923

[6] D. Seebach, R. Imwinkelried, T. Weber): in R. Scheffold (Ed.) "*Modern Synthetic Methods*", Vol.4, Berlin (**1986**)

[7] a) L. Vargha, E. Kasztreiner, *Chem. Ber.* **1959** *92*, 2506
b) C. Morphain, M. Tisserand, *J. Chem. Soc. Perk. Trans. I*, **1979** 1379
c) Y. Merrer, A. Dureault, C. Gravier, D. Languin,J.C. Depezay, *Tet. Lett.* **1985** *26*, 319

[8] J.H. Weiss,W.C. Christian, *Neuron*, **1990** *3*, 321 ; *Brain-Research* **1988** *191*, 441

[9] a) T. Teshima, S. Nomoto, T. Wakamiya, T. Shiba, *Bull. Chem. Soc. Jpn.* **1977** *50*, 3372;
b) P.J. Dunn,R. Häner,H. Rapoport, *J. Org. Chem.* **1990** *55*, 5017;
c) G. Cardillo, M. Orena, P. Maurizio, S. Sandri, C. Tomasini, *Tetrahedron* **1991** *47*, 2263;
d) L.D. Arnold, R.G. May, J.C. Vederas, *J. Am. Chem. Soc.* **1988** *110*, 2237;
e) K. Nacajima, T. Tanaka, K. Morita, K. Okawa, *Bull. Chem. Soc. Jpn.* **1980** *53*, 283;
f) N. Shigematsu, H. Setoi, I. Uchida,T. Shibata,H. Terano,M. Hashimoto, *Tet. Lett.* **1988** *29*, 5147 ;
g) U. Schmidt, K. Mundinger, B. Riedl, G. Haas, R. Lau; *Synthesis* **1992** 1201;
h) P.G. Mattingly, M.J. Miller, *J. Org. Chem.***1980** *45*, 410;
i) H. Estermann, D. Seebach, *Helv. Chim. Acta* **1988** *71*, 1824

[10] R.W. Hoffmann; *Chem. Reviews* **1989** *89*, 1841; *Angew. Chem.* **1992** *104*, 1147

[11] J. Mulzer, R. Becker, E. Brunner, *J. Am. Chem. Soc.* **1989** *111*, 7500
[12] a) S.H. Snyder, Drogen der Psyche, Seite 155, Heidelberg (**1988**); b) V.J. MerLuzzi; *Science* **1990** *250*, 1411; c) R. Pauwels, *Nature* **1991** *343*, 470; d) B.E. Evans, M.G. Bock, K.E. Rittle, R.M. DiPardo, W.L. Whitter, D.F. Veber,P.S. Anderson, R.M. Freidinger; *Proc. Natl. Acad. Sci. U.S.A.* **1986** *83*, 4918; e) B.E. Evans; *J. Med. Chem.* **1988** *31,* 2235
[13] L. Sternbach; *J. Org. Chem.* **1962** *27*, 3788
[14] a) B.D. Christie, H. Rapoport; *J. Org. Chem.* 1985 *51*, 1239;
b) A.M.P. Koskinen, H. Rapoport; *J. Org. Chem.* **1989** *54*, 1859

Enzymemimetic C-C and C-N Bond Formations

D. ENDERS

Institut für Organische Chemie, Rheinisch-Westfälische Technische Hochschule,
Professor-Pirlet-Str. 1, D-52074 Aachen

Summary. Simple C2 and C3 building blocks, such as dihydroxyacetone phosphate (DHAP), phosphoenol pyruvate (PEP) and the active acetaldehyde or glycolaldehyde, are used by nature in enzyme catalyzed, highly stereoselective C-C bond forming reactions. The first enantiopure chemical synthetic equivalents of these building blocks are synthesized and employed in enzymemimetic C-C bond formations. Stereoselective C-N bond formations can be mimicked by using a chiral ammonia equivalent in Michael additions to enoates. It turns out, that this biomimetic strategy opens a way to both enantiomers of the final products in a broad applicability and that diastereo- and enantioselectivities are reached, which compare well with the corresponding enzymatic processes.

1. Introduction

It has always been the dream of chemists to synthesize complex organic molecules in a similar manner, elegance and efficiency as mother nature does it. For a long time, biomimetic strategies have therefore been used successfully in organic synthesis. However, it is surprising that for the transfer of simple C2 and C3 building blocks, such as for instance dihydroxyacetone phosphate (DHAP) **A**, phosphoenol pyruvate (PEP) **B**, the "active acetaldehyde" **C** and the "active glycolaldehyde" **D**, chemical synthetic equivalents are barely available. This is particularly true when one attempts to mimic the high stereoselectivities of the naturally occurring enzyme-catalyzed reactions, namely to transfer the synthons **A-D** diastereo- and enantioselectively to an electrophilic substrate by asymmetric synthesis [1].

We would like to report here our first advances in the development of new synthetic methods employing chirally modified organometallic reagents, which can be used as synthetic equivalents of the building blocks **A-D** and their appli-

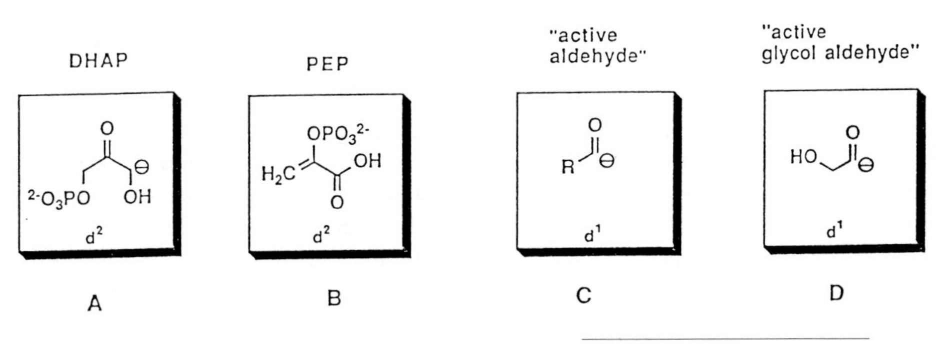

Stereoselective Synthesis
Editors: Ottow, Schöllkopf, Schulz

cation in diastereo- and enantioselective syntheses of potentially biologically active compounds.

2. Enzymemimetic C-C Bond Formations

2.1. Diastereo- and Enantioselective C-C Bond Formations with a Chiral Dihydroxyacetone phosphate Equivalent

Dihydroxyacetone phosphate (DHAP) **A** is used in nature as a C3-methylene component in enzyme catalyzed aldol reactions to form 2-ketoses. Recently, several research groups have successfully applied this natural synthetic principle by using aldolases as catalysts and a wide variety of aldehyde carbonyl components to form carbohydrates and related structures [2-12]. A chemical alternative is shown in scheme 1. 2,2-Dimethyl-1,3-dioxan-5-one (**1**), which can be made in molar quantities from the simple starting materials nitromethane, formaldehyde and acetone in 4 steps, can be used as a DHAP-enolate equivalent, utilising the SAMP-/RAMP-hydrazone method [13]. After metalation of the hydrazone (S)-**2** and trapping with a large number of alkyl halides, followed by oxidative cleavage of the resulting products (S,S)-**3**, the ketones (S)-**4** are obtained in high enantiomeric purity. By simply changing the auxiliary from SAMP to RAMP, the corresponding enantiomers can also be synthesized [14].

If commercially available, protected polyoxygenated iodides are used as electrophiles, an efficient and stereochemically flexible entry to deoxysugars is opened. In this way, all theoretically possible stereoisomers of the protected deoxysugars shown in scheme 2 can be synthesized [15].

In the really enzymemimetic process, the aldol reaction of (S)-**2** with aldehydes, one obtains as depicted in scheme 3 polyhydroxylated ketones of the anti-(3S,4S)-configuration, hitherto difficult to prepare by either chemical or enzymatic methods (D-tagatose-1,6-diphosphate aldolase) [9]. Whereas excellent diastereomeric excess could be obtained after flash-chromatography (de $\geq$ 95%), the maximum enantiomeric excess of so far up to 82% has yet to be optimized [16].

Through chemoselective esterification of the primary hydroxy group of the 2-ketoses to the corresponding (S)-3,3,3-trifluoro-2-methoxy-2-phenylpropionic acid [(S)-MTPA] esters [17], we opened an easy and reliable way to determine the diastereomeric and enantiomeric excess by ^{13}C NMR spectroscopy. In addition, by correlation with our SAMP-/RAMP-hydrazone method, the absolute configurations can be assigned [16].

The FDP-aldolase catalyzed DHAP-addition to glyoxylic acid, examined in detail by Wandrey and Bossow-Berke under continuous flow reaction conditions [18], required the determination of the relative and absolute configuration. Employing the SAMP-/RAMP-hydrazone method, the aldol reaction of **1** with benzyl glyoxylate afforded a doubly protected adduct, which was deprotected with trifluoroacetic acid anhydride (TFAA) and subsequent debenzylation with H_2/Pd-C to give the final product, identical with the enzyme product. The independent "chemical" synthesis and the enzymatic route are compared in scheme 4. It

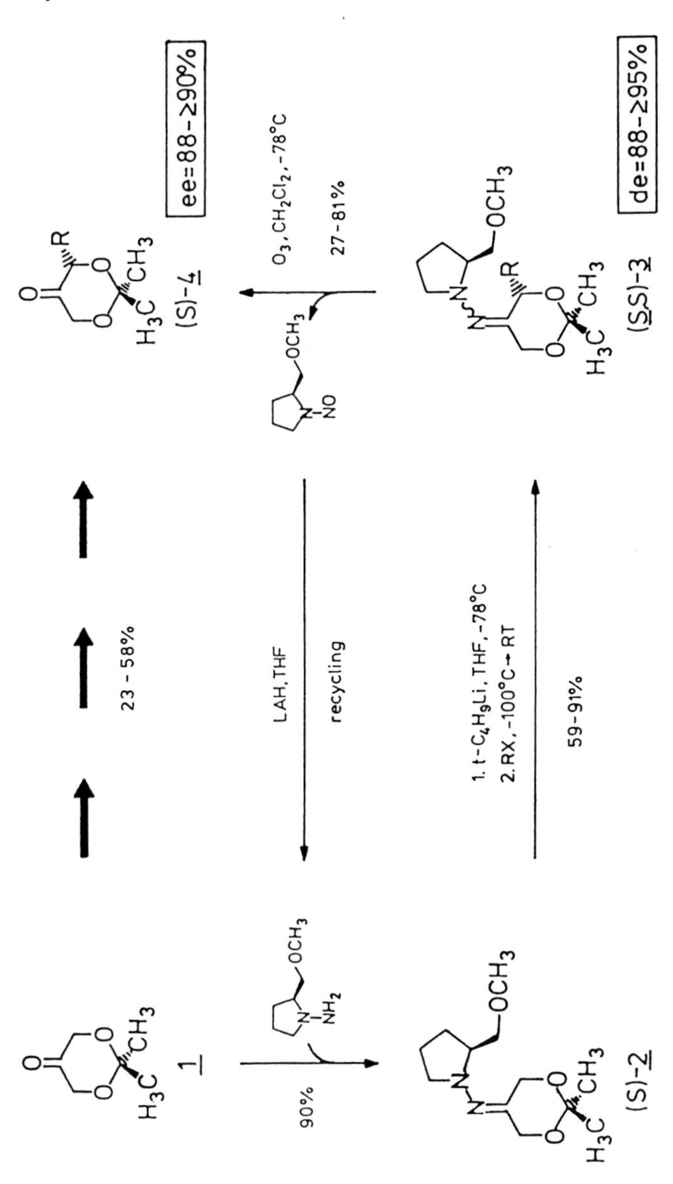

R= CH_3, C_2H_5, n-C_6H_{13}, $CH(CH_3)_2$, $CH_2CH{=}CH_2$, $CH_2C_6H_5$, CH_2OBn, $(CH_2)_2OBn$, CH_2COOCH_3

Scheme 1. Enantioselective alkylations with a novel DHAP-equivalent

SAMP

1. t-BuLi, THF, -78°C

2. (R)-5 -100°C -> r t

O_3, CH_2Cl_2, -78°C

(S)-2

3

1

53%

(S,S)-4

de = 94%
ee ≥ 99%

all 4 stereoisomers available (RAMP, (S)-5)

Scheme 2. Stereoselective *de novo* - synthesis of deoxysugars

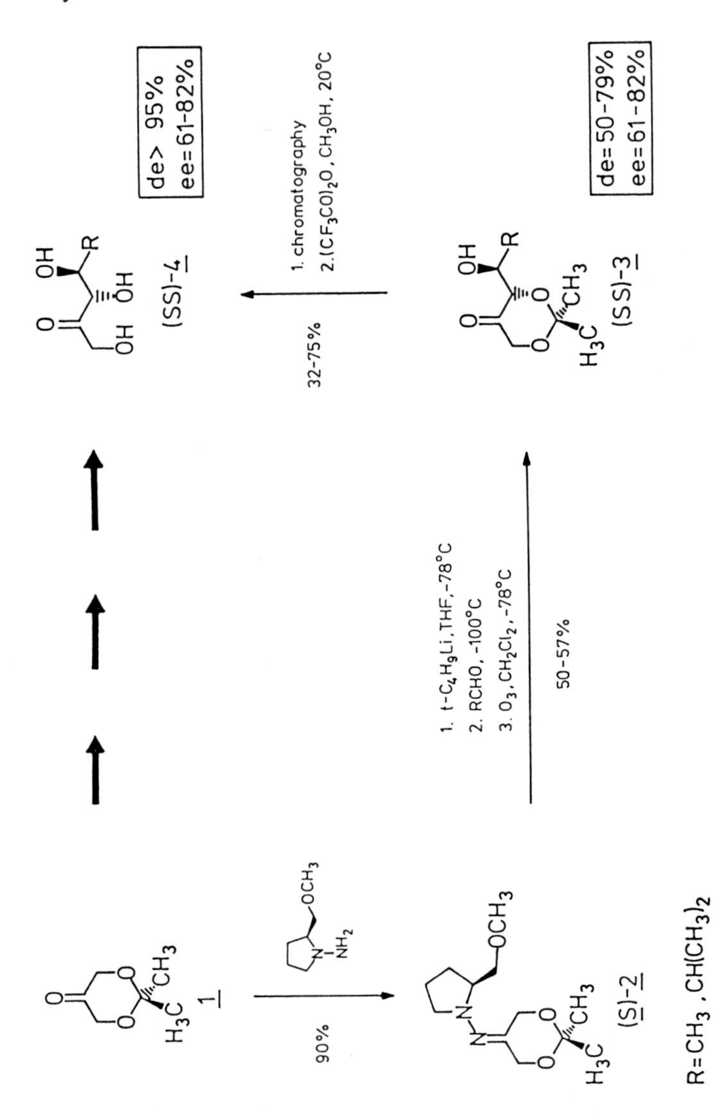

Scheme 3. Diastereo- and enantioselective aldol reactions with a DHAP- equivalent

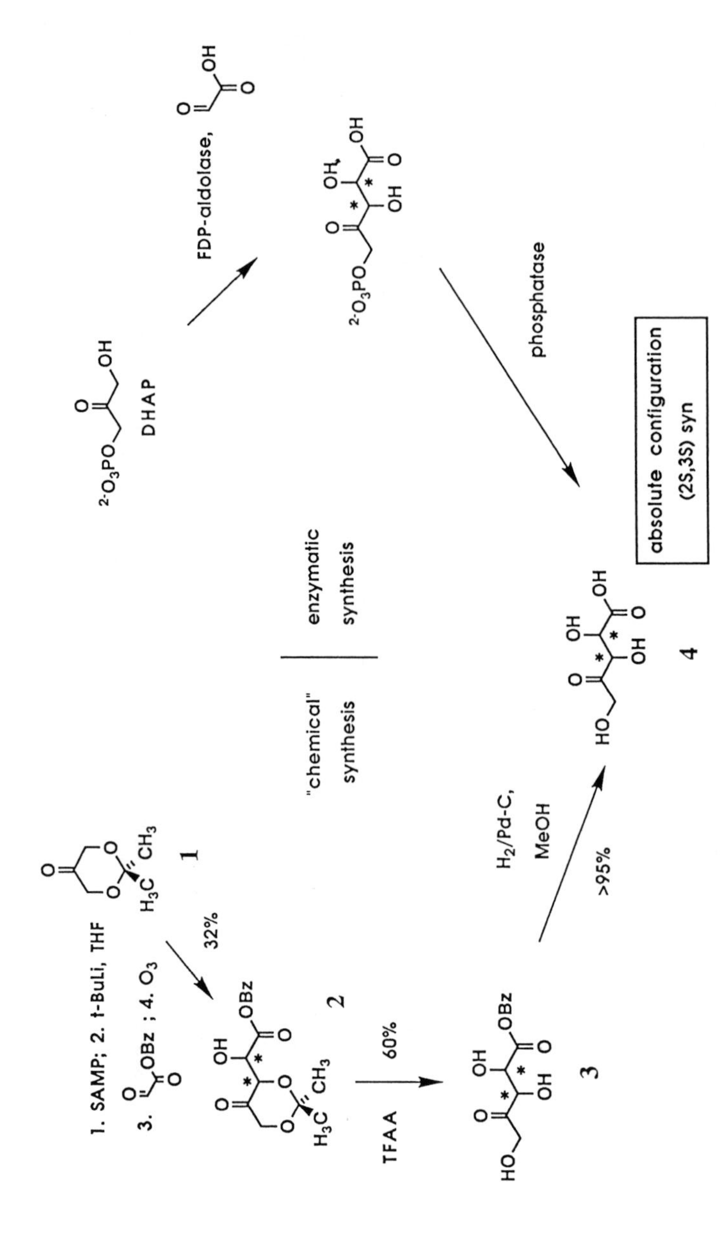

Scheme 4. Determination of relative and absolute configuration by independent asymmetric synthesis

turns out that the syn (2S,3S) configuration is obtained and thus in accordance with the proposed mechanism even with the relatively "unnatural" benzyl glyoxylate substrate the syn stereochemistry common for all FDP-aldolase catalyzed reactions [15].

The remarkable synthetic flexibility of the new enzymemimetic processes to transfer the dihydroxyacetone C3-unit to electrophiles can be demonstrated by using two different electrophiles to effect α,α'-functionalisation. Scheme 5 shows some examples of polyfunctional molecules of high diastereo- and enantiomeric purity, which can easily be made in gram quantities in 4 steps and good overall yields [15].

Since the pioneering work of Kagan in 1972 on catalytic asymmetric hydrogenations using the diphosphine ligand DIOP [19], C_2-symmetric chiral auxiliaries have proven to be particularly useful in asymmetric synthesis [20]. If the same electrophile is used twice in our α,α'-bisalkylations of dihydroxyacetone acetonide **1**, one obtains with virtually complete asymmetric induction C2-symmetric ketones (de, ee $\geq$ 98%) [21]. Obviously, the second alkylation occurs regioselectively at the α'-position without epimerization of the α-stereogenic center and the uniform SAMP-hydrazone mechanism of electrophilic substitution α to the carbonyl group [22, 23] is not disturbed by the preexisting α-stereocenter (scheme 6).

A first application of the new, very efficient asymmetric synthesis of C_2-symmetric ketones is described in scheme 7. Since the Center for Disease Control in Atlanta (USA) defined the diagnostic term AIDS (Acquired Immunodeficiency Syndrome) in 1982 [24] only three medications have been authorized for treatment of AIDS: 3'-azido-3'-deoxythymidine (AZT, Wellcome, 1987), 2',3'-dideoxyinosine (DDI, Bristol Myers Squibb, 1992), and 2',3'-dideoxycytosine (DDC, Hoffmann La Roche, 1992), which was recently introduced for limited use. These drugs inhibit the enzyme reverse transcriptase of the human immunodeficiency virus (HIV). Nevertheless, they are only able to prolong somewhat the survival of patients with advanced cases of AIDS. They also lead to considerable side-effects (bone marrow damage, neuropathy) and to the generation of more resistant strains of the virus [25].

Since the elucidation of its structure in 1989 HIV-1 protease has become a new, highly favored target for chemotherapy [26]. This protease belongs to the class of acidic aspartate proteases and has an unusual homodimeric C_2-symmetrical structure. Starting from the Phe-Pro cleavage point most frequently preferred by the enzyme, Erickson and Kempf et al. [27] developed the C_2-symmetrical, highly selective HIV-1 protease inhibitor A-74704 (scheme 7). They demonstrated that the marked inhibiting effect of the inhibitor relies on its optimal fit in the active center of the C_2-symmetrical protease.

According to our protocol, the α,α'-bisbenzylation of **1** afforded the corresponding C_2-symmetric ketone with practically complete asymmetric induction. In 6 further steps, using well known basic transformations and a Mitsunobu reaction to introduce the nitrogen functionalities, we reached the diastereo- and enantiomerically pure diamino alcohol **6**, which is easily transformed to A-74704 according to Kempf et al. (Abbott) and Dreyer et al. (SmithKline Beecham)

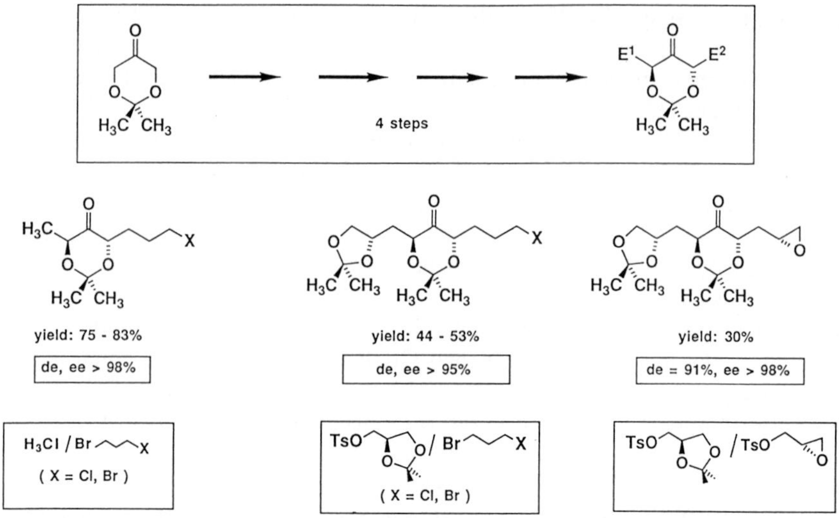

Scheme 5. Diastereo- and enantioselective synthesis of protected, polyfunctional ketodioles from dihydroxyacetone

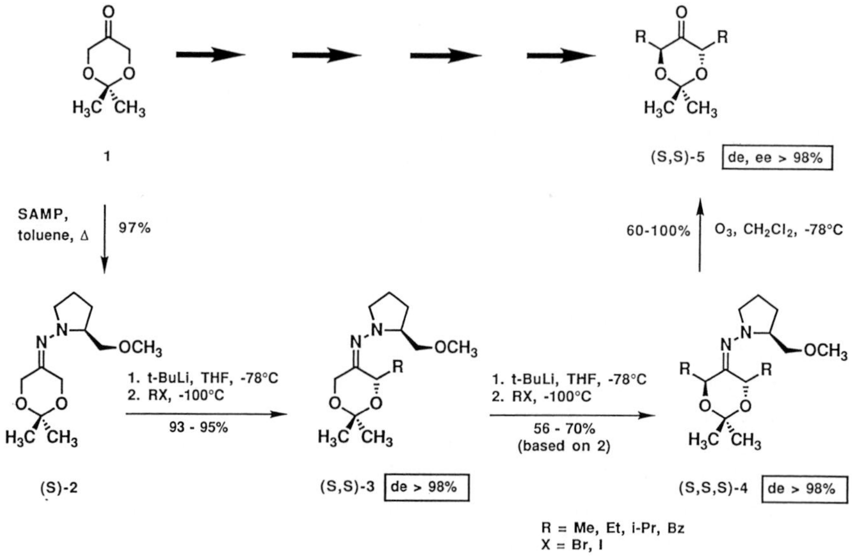

Scheme 6. Diastereo- and enantioselective synthesis of protected, C_2- symmetric ketodiols using the SAMP-/ RAMP-hydrazone method

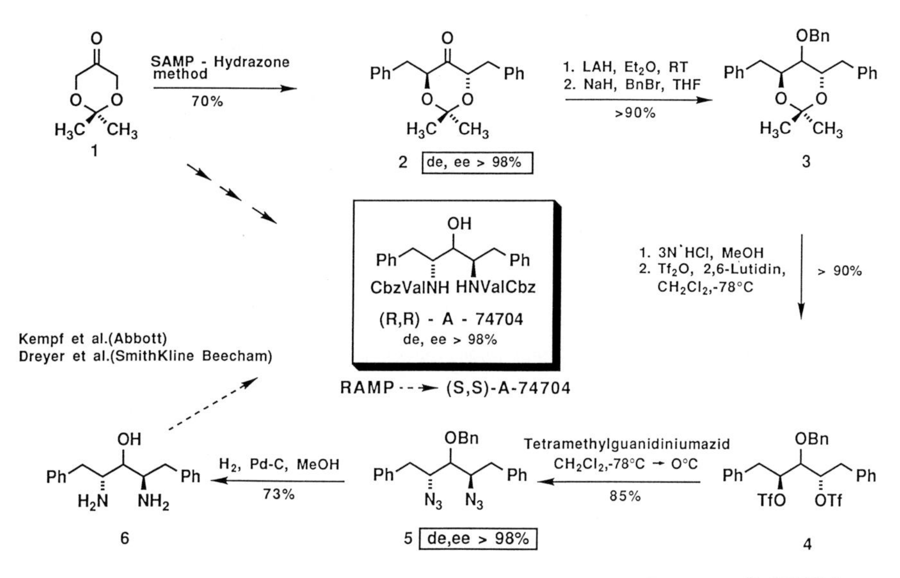

Scheme 7. Diastereo- and enantioselective synthesis of A-74704, a C_2-symmetrical HIV-1 protease inhibitor [28]

(scheme 7) [28]. It is interesting to note that **6** is also the key building block of a HIV-1 protease inhibitor developed by Bayer AG and Hoechst AG [29].

The asymmetric synthesis described in scheme 7 provides an efficient and stereochemically flexible entry to C_2-symmetric HIV-1 protease inhibitors. The inhibiting effect and the pharmacological properties of these new agents may possibly be improved further for the treatment of AIDS, as certain properties can be varied almost at will: the stereogenic centers, which can be selected by the choice of the auxiliary (SAMP/ RAMP); the side chain (here $PhCH_2$), by the choice of electrophile; and the substituents on the amino group.

The products presented in scheme 5 show that it should be possible by subsequent introduction of amino groups through halogen or oxirane functionalities, to get an efficient entry into a series of polyhydroxylated piperidines (azasugars) [6,7,30] by intramolecular reductive amination. This class of compounds gained much interest recently because of its biological activity as potential glycosidase inhibitors [31-33]. In scheme 8 novel substitution patterns of polyhydroxy piperidines (boxed), which we synthesized stereoselectively by our own new procedures [34], are compared with known compounds derived from natural sources.

Our synthetic approach is demonstrated by the two typical examples shown in schemes 9 and 10. In the oxirane azide ring opening variant piperidine derivatives bearing 4 stereogenic centers are formed (scheme 9). Whereas the stereocenter α to nitrogen in the latter is not fully controlled and a mixture of C2-epimers is obtained, the bromide azide substitution variant depicted in scheme 10 (stereotriade) works with virtually complete asymmetric induction and only one stereoisomer is formed (de,ee ≥ 98%).

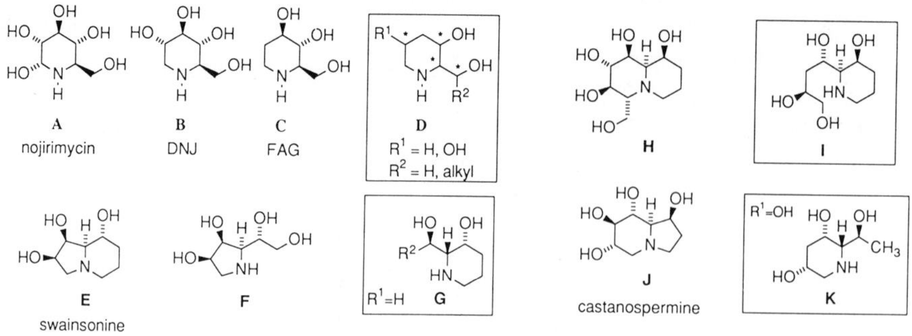

Scheme 8. Novel polyhydroxylated piperidines (azasugars) as potential glycosidase inhibitors [34]

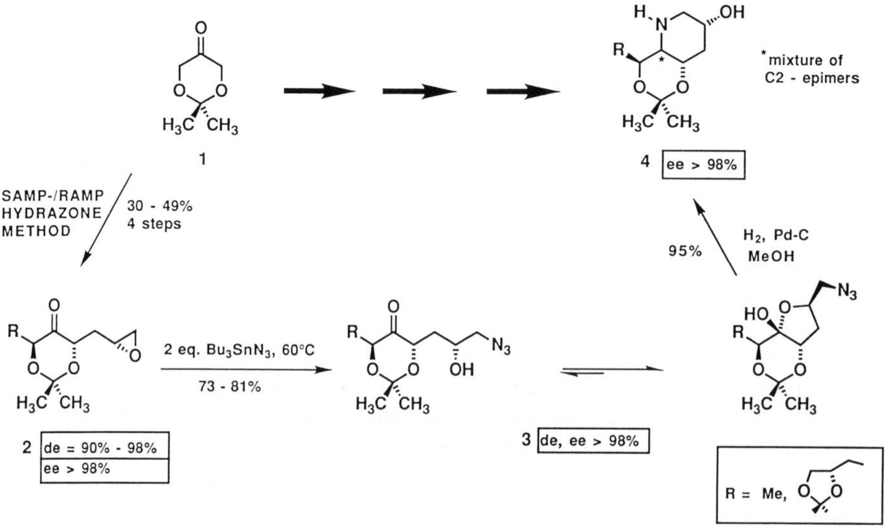

Scheme 9. Flexible, diastereo- and enantioselective synthesis of novel polyhydroxylated piperidines (oxirane azide ring opening variant) [15]

In summary, the procedures described above offer an efficient entry to novel polyhydroxylated piperidine structures of excellent diastereo- and enantiomeric purities and in good overall yields. The absolute configurations of the stereogenic centers can be selected by choosing the right auxiliary (SAMP, RAMP). By variation of the first electrophile [21] (R^2 in **D**) a broad range of side chains may be introduced. In addition, an extension to different ring sizes seems possible by changing the length of the second electrophile.

Finally, it should be mentioned that the chiral DHAP equivalent (S)-**2** (scheme 1) reacts after metalation with Michael acceptors, such as enoates or α,β-unsaturated phosphonates, under 1,4-addition. The resulting 3-substituted 4,6-dihydroxy-5-oxoesters and phosphonates are obtained in good overall yields and with excellent diastereo- and enantiomeric excess [35].

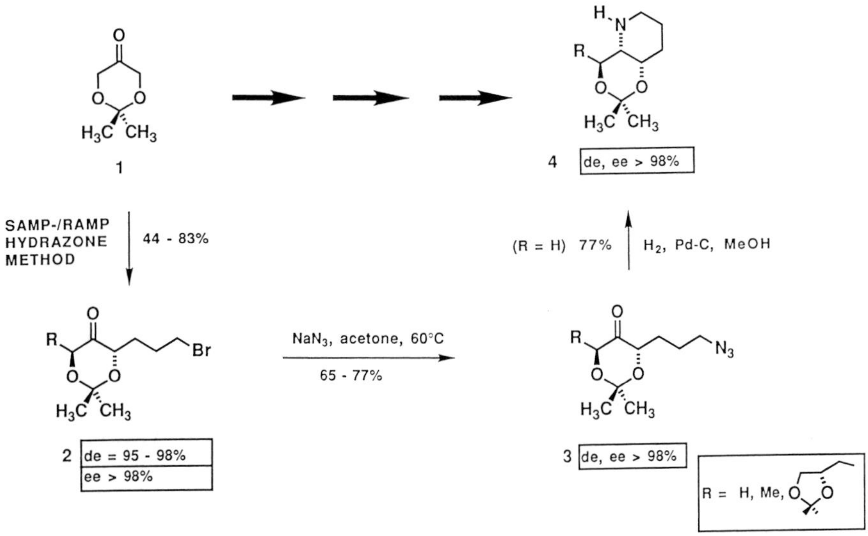

Scheme 10. Flexible, diastereo- and enantioselective synthesis of novel polyhydroxylated piperidines (bromide azide substitution variant)

2.2 Diastereo- and Enantioselective C-C Bond Formations with a Chiral Phosphoenolpyruvate Equivalent

The transfer of the pyruvate unit as phosphoenolpyruvate (PEP) to aldoses results in 4-hydroxy-2-oxocarboxylic acid structures and constitutes an important natural C-C bond forming reaction for the biosynthesis of ulosonic [36] and sialic acids [37,38]. Important compounds such as N-acetylneuraminic acid, 3-deoxy-D-*manno*-octulosonic acid (KDO), and 3-deoxy-D-*arabino*-2-heptulosonic acid 7-phosphate (DAHP), the precursor of shikimic acid, are formed in this way.

Whereas DHAP-aldolase catalyzed aldol reactions have been thoroughly investigated [2-12], the use of pyruvate-aldolases is still in its infancy [39,40]. As the first chemical synthetic equivalents of PEP, which are able to transfer the pyruvic acid d^2-synthon (boxed), we have synthesized chiral pyruvate hydrazones from simple chemicals and (S)-proline (scheme 11).

Our initial attempts at metalating hydrazones of methyl and *tert*-butyl pyruvate and trapping them with electrophiles were unsuccessful and led primarily to self-acylated products (scheme 12). Only after we sterically blocked the ester reactivity as the 2,6-di-*tert*-butyl-4-methoxyphenyl ester was the direct manipulation of the corresponding azaenolates possible.

The synthesis of the sterically hindered 2-ketoesters succeeded in a simple fashion and in excellent yields by the esterification of ethyl oxalyl chloride with the lithium salt of 2,6-di-*tert*-butyl-4-methoxyphenol followed by the chemoselective nucleophilic addition of methyl or ethyl Grignard reagents to the unsymmetrical ethyl aryl oxalate. Reaction with (S)-1-amino-2-(methoxymethyl)pyrro-

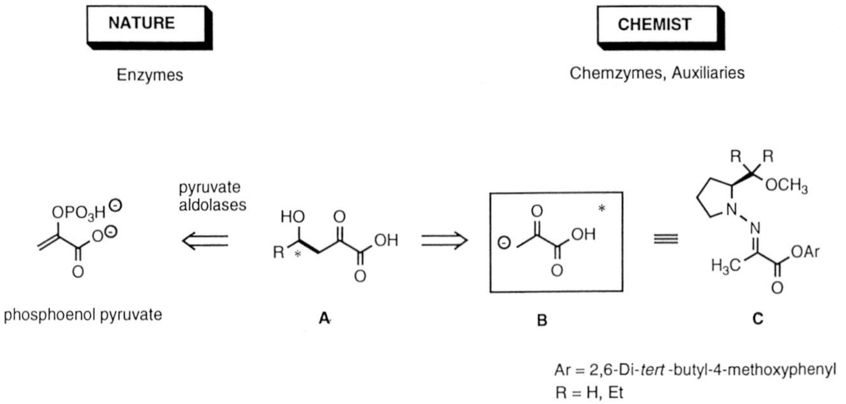

Ar = 2,6-Di-*tert*-butyl-4-methoxyphenyl
R = H, Et

Scheme 11. Enzyme catalyzed PEP-transfer and chemical mimicry via metalated chiral hydrazones

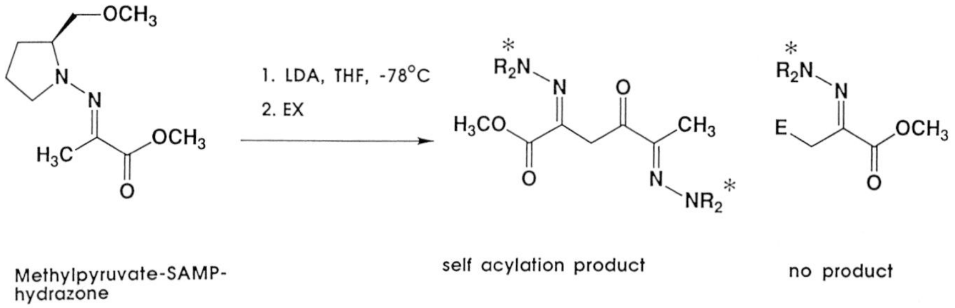

Scheme 12. First test reactions using methylpyruvate SAMP-hydrazone as chiral PEP equivalent

lidine (SAMP) finally gave the aryl pyruvate SAMP-hydrazone and its higher homologue as pale yellow solids in almost quantitative yield [41] (scheme 13).

The new hydrazones were first tested in simple α-alkylation reactions. As is shown in scheme 14, metalation with Lochmann-Schlosser base in THF at low temperature yielded highly reactive azaenolates, which were alkylated by a number of alkyl halides at -100°C in good yields and with high diastereomeric excess (de = 85 - ≥ 95%). It is noteworthy that, besides the usual oxidative cleavage with ozone, the SAMP-hydrazones could be converted without racemization to the final 3-substituted 2-oxoesters under very mild conditions with boron trifluoride-ether in acetone/water with the addition of paraformaldehyde. Thus, this method permitted for the first time the highly enantioselective transfer of a homologous pyruvate unit to electrophiles (scheme 14) [41].

The absolute configuration of the product hydrazones and thus also of the ketoesters was determined by X-ray structure analysis of a crystalline benzylated SAMP-hydrazone. The (S) configuration found at the newly formed stereogenic center is again in agreement with that predicted by the postulated mechanism for

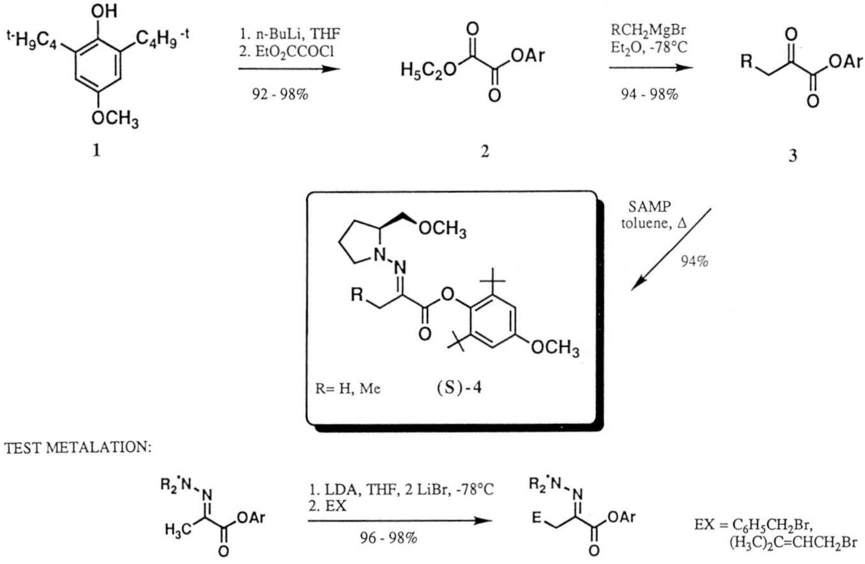

Scheme 13. Synthesis and first reactions of sterically hindered 2-ketoester SAMP-hydrazones as chiral PEP-equivalents

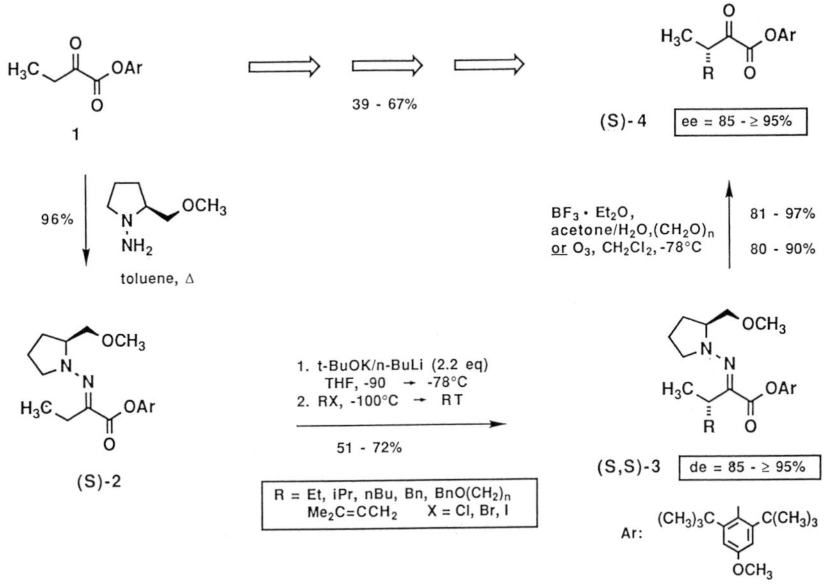

Scheme 14. Enantioselective synthesis of 3-substituted 2-ketoesters

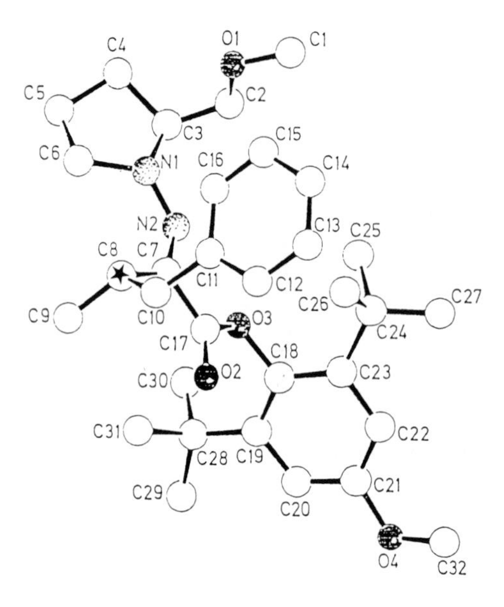

Figure 1. Structure of the SAMP-hydrazone (S,S)-**3** (R=Bn) in the crystal (Schakal plot)

electrophilic substitutions via SAMP-/RAMP-hydrazones [22,23] (figure 1). The crystal structure nicely demonstrates that the ester carbonyl group is blocked by the two bulky *tert*-butyl groups.

As a first extension of the new overall enantioselective α-alkylation of 2-ketoesters we developed an efficient asymmetric synthesis of 3-substituted cyclic hemiketals of ω-hydroxy-2-oxoesters [42]. As is depicted in scheme 15, the metalated SAMP-hydrazones were trapped with a number of O-benzyl- and O-silyl-protected ω-hydroxy-1-iodoalkanes, which after hydrazone cleavage and deprotection afforded the cyclic hemiketals (de, ee ≥ 98%), structurally modified deoxygenated analogs of ulosonic acids. To our knowledge only two methods have been reported for the synthesis of similar simple tetrahydropyran derivatives [43, 44].

The relative and absolute configurations shown are based on Nuclear Overhauser experiments and an X-ray crystal structure analysis shown in Figure 2.

The true baptism of fire came when we tried to perform stereoselective aldol reactions with a broad range of aldehydes as electrophiles - the real biomimetic reaction sequence. As scheme 16 shows, we were pleased to see that indeed highly overall enantioselective aldol reactions could be performed and benzyloxymethyl (BOM) protected 4-hydroxy-2-oxocarboxylic acid esters of high enantiomeric purity (ee ≥ 98%) were isolated starting with aryl pyruvates in good overall yields [45]. In order to get such very high asymmetric inductions it was necessary to exchange SAMP for the sterically more demanding auxiliary (S)-1-amino-2-(1-ethyl-1-methoxypropyl)pyrrolidine (SAEP) [46] and to metalate the corresponding SAEP-hydrazone with lithium diisopropylamide (LDA) in the presence of two equivalents of lithium bromide.

H_3C OAr (O, O) **1** ⟹ ⟹ ⟹ ⟹ 40 - 54 % OH CO_2Ar CH_3 (S,S)-**5** ee = de ≥ 98 %

96 % N-NH_2, OCH_3

1. H_2, Pd/C, EtOH ($P = PhCH_2$) or $BF_3 \cdot Et_2O$, acetone/ H_2O ($P = t BuMe_2Si$)
2. Chromatography
95 - 100 %

(S)-**2** — 1. tBuOK/nBuLi, THF, -90 - -78°C 2. $PO(CH_2)_nI$, -100°C; 55 - 67 % → (S,S)-**3** de = 85 - ≥ 98 % — O_3, CH_2Cl_2, -78°C; 80 - 86 % → (S)-**4** ee = 85 - ≥ 98 %

(n = 2 - 4)
Ar = 2,6 - Di -t - butyl - 4 - methoxyphenyl

Scheme 15. Diastereo- and enantioselective synthesis of structurally modified deoxygenated analogs of ulosonic acids [42]

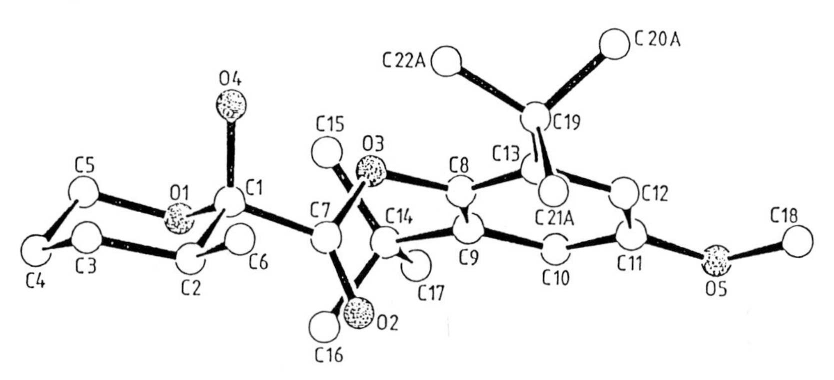

Figure 2. Determination of the configuration of 7 (n = 3) by X-ray structure analysis

In this manner, for instance, the protected natural product 2-keto-3-deoxygluconate (KDG) [39] starting from isopropylidene-protected (R)-glyceraldehyde was obtained in diastereo- and enantiomerically pure form. The absolute configurations given are based again on a crystal structure analysis of a single crystal of a hydrazone adduct ($R = Me_2CHCH_2$, figure 3) and by the assumption of a uniform reaction mechanism. The relative topicity found is in agreement with the stereochemical outcome previously observed in aldol reactions with SAMP/RAMP-hydrazones [47, 48, 23]. Beside the synthetic value of the highly enantioselective and generally applicable PEP-transfer, it is especially noteworthy that virtually complete asymmetric inductions could be obtained with a methyl ketone as methylene component, a stereochemical situation notoriously difficult to control in the aldol reaction ("acetate problem") [49]. In 1978 we re-

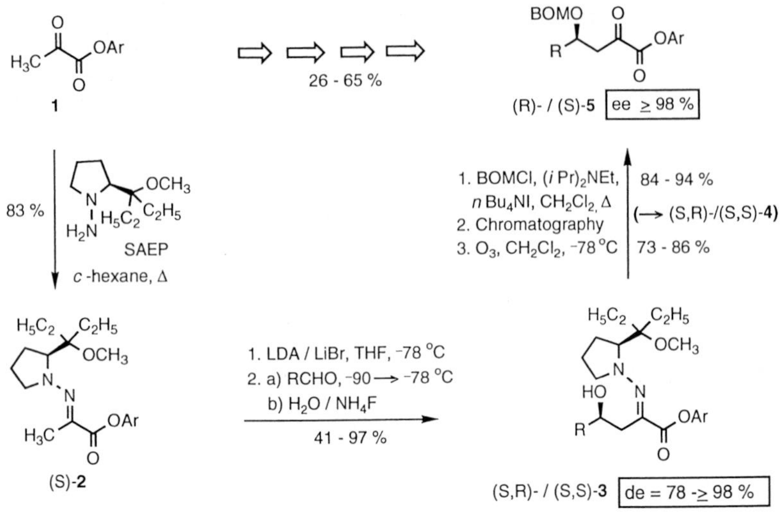

H_3C OAr O O
1
26 - 65 %
BOMO O R OAr O
(R)- / (S)-**5** ee ≥ 98 %

83 %
OCH_3 N H_5C_2 C_2H_5 H_2N
SAEP
c-hexane, Δ

1. BOMCl, (*i* Pr)$_2$NEt, *n* Bu$_4$NI, CH_2Cl_2, Δ
2. Chromatography
3. O_3, CH_2Cl_2, -78 °C
84 - 94 %
(⟶ (S,R)-/(S,S)-**4**)
73 - 86 %

H_5C_2 C_2H_5 OCH_3 N N H_3C OAr O
(S)-**2**

1. LDA / LiBr, THF, -78 °C
2. a) RCHO, -90 ⟶ -78 °C
b) H_2O / NH_4F
41 - 97 %

H_5C_2 C_2H_5 OCH_3 N HO N R OAr O
(S,R)- / (S,S)-**3** de = 78 - ≥ 98 %

Scheme 16. Enantioselective aldol reactions with the first chiral PEP-equivalent [45]

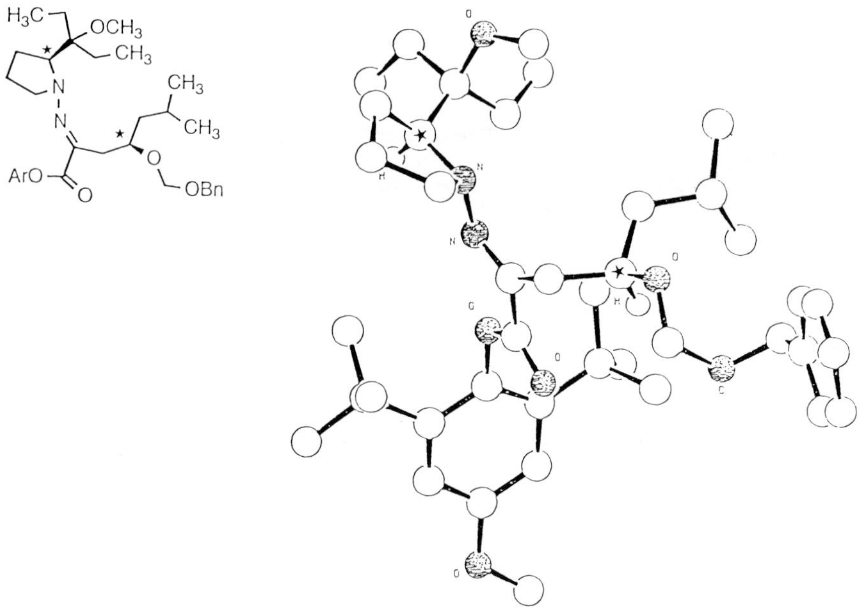

H_3C OCH_3 CH_3 N N CH_3 CH_3 ArO O O OBn

Figure 3. Determination of absolute configuration by X-ray structure analysis

ported the first enantioselective aldol reaction employing our SAMP/RAMP hydrazone method, but obtained only mediocre ee-values. Although we developed in the meantime aldol procedures resulting in excellent de- and ee-values (crystallization trick [23, 50], α-silylketone enolates [51]), it took us almost 15 years to find conditions for the complete asymmetric induction in aldol reactions via hydrazone azaenolates.

2.3 Asymmetric Nucleophilic Acylations with Synthetic Equivalents for the "Active Aldehydes"

Fascinating for the synthetic chemist is the ability of nature to transfer aldehydes in the sense of a nucleophilic acylation to other aldehydes. This occurs by thiamine pyrophosphate catalysis via the so-called "active acetaldehyde" (scheme 17) and "active glycolaldehyde". Inspired by these biochemical pathways and because we thought the time was ripe, we initiated a programme directed towards the development of asymmetric nucleophilic acylation reactions in the late seventies [52-55]. In our first attempts we investigated acyloin type C-C-bond formations using metalated chiral formamides, thioformamides and aminonitriles bearing mainly (S)-proline derived chiral auxiliaries. Recently we made a breakthrough by utilizing metalated aminonitriles based on a novel aminodiol auxiliary and prepared from the so-called wrong enantiomer of the chloramphenicol synthesis (Boehringer Mannheim company) (scheme 18) [56]. The new method for asymmetric conjugated nucleophilic acylation allows the transfer of aldehydes to enoates resulting in 3-substituted 4-oxoesters of high enantiomeric purity (ee = 72 - > 95%). It should be mentioned that the chiral auxiliary can be recycled in 91% yield after the $CuSO_4$-mediated aminonitrile hydrolysis. Cyclohexenone can also be used successfully as Michael acceptor, which constitues a new asymmetric synthesis of 1,4-diketones with excellent overall yields (51 - 95%) for the 3 step procedure and enantiomeric excess of 70 - 98% (scheme 19) [57].

When we used α,β-unsaturated sulphones and phosphonates as well as aldehydes (acyloin reaction) as electrophiles to trap the metalated aminonitriles, moderate to good ee-values were obtained and we are still engaged to optimize these variants [58].

Although the chiral auxiliary employed in the stoichiometric aminonitrile chemistry described above can be recycled in good chemical yield and maintaining its optical purity, of course, a more elegant and economical solution would be a catalytic process. Therefore we tried to imitate nature again (scheme 17) by developing an enantioselective thiazolium salt catalysis.

First of all, we had to work out a simple and flexible method for the synthesis of chirally modified thiazolium salts, preferably with the chiral group attached to the nitrogen atom. With a slightly modified procedure already published by Tagaki et al. [59], we were able to prepare a number of novel chiral thiazolium salts.

By exchange of the gegenion of the hygroscopic chlorides to tetrafluoroborates or hexafluorophosphates, crystalline, non hygroscopic salts were obtained,

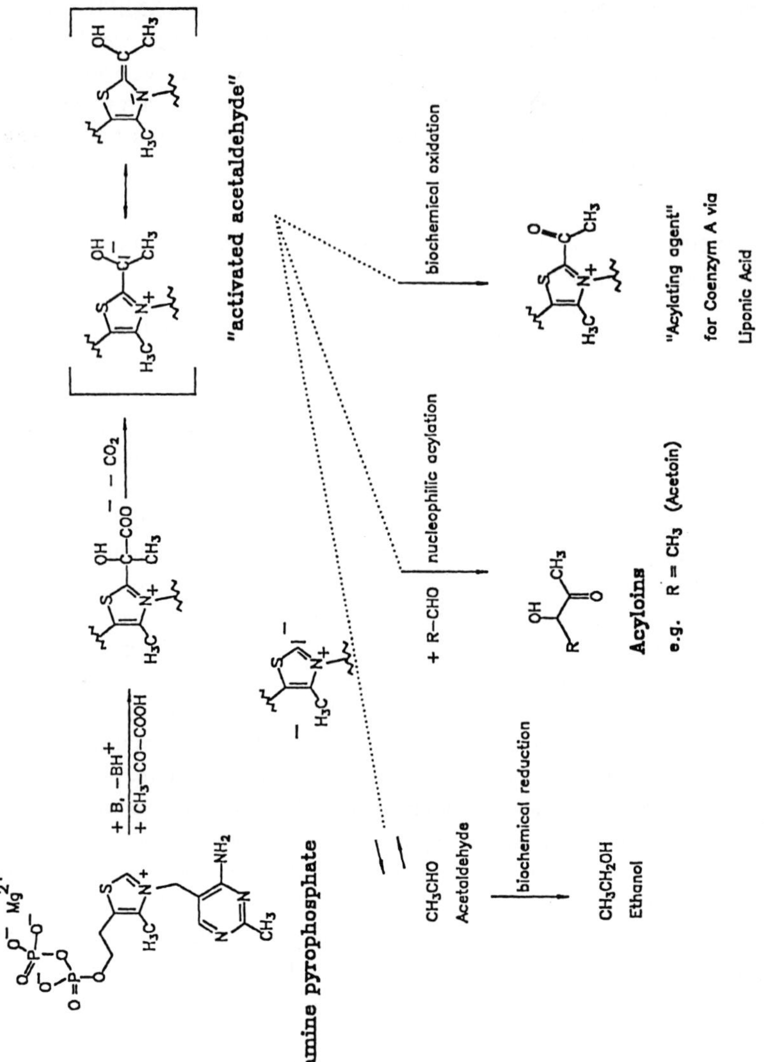

Thiamine pyrophosphate

$+ B, -BH^+$

$+ CH_3-CO-COOH$

$- CO_2$

"activated acetaldehyde"

CH_3CHO

Acetaldehyde

biochemical reduction

CH_3CH_2OH

Ethanol

$+ R-CHO$

nucleophilic acylation

Acyloins

e.g. $R = CH_3$ (Acetoin)

biochemical oxidation

"Acylating agent"

for Coenzym A via

Liponic Acid

Scheme 17. Thiamine pyrophosphate-catalyzed reactions in nature

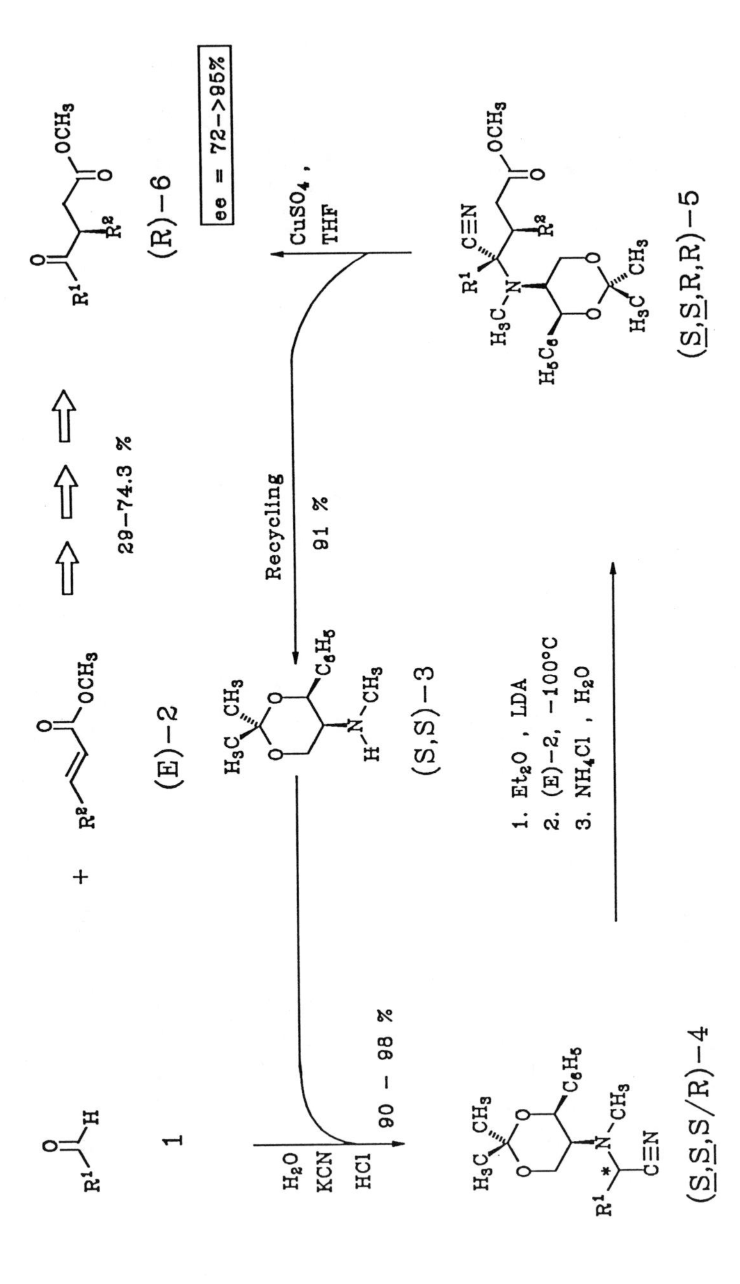

Scheme 18. Enantioselective synthesis of 3-substituted 4-oxoesters via metalated chiral aminonitriles

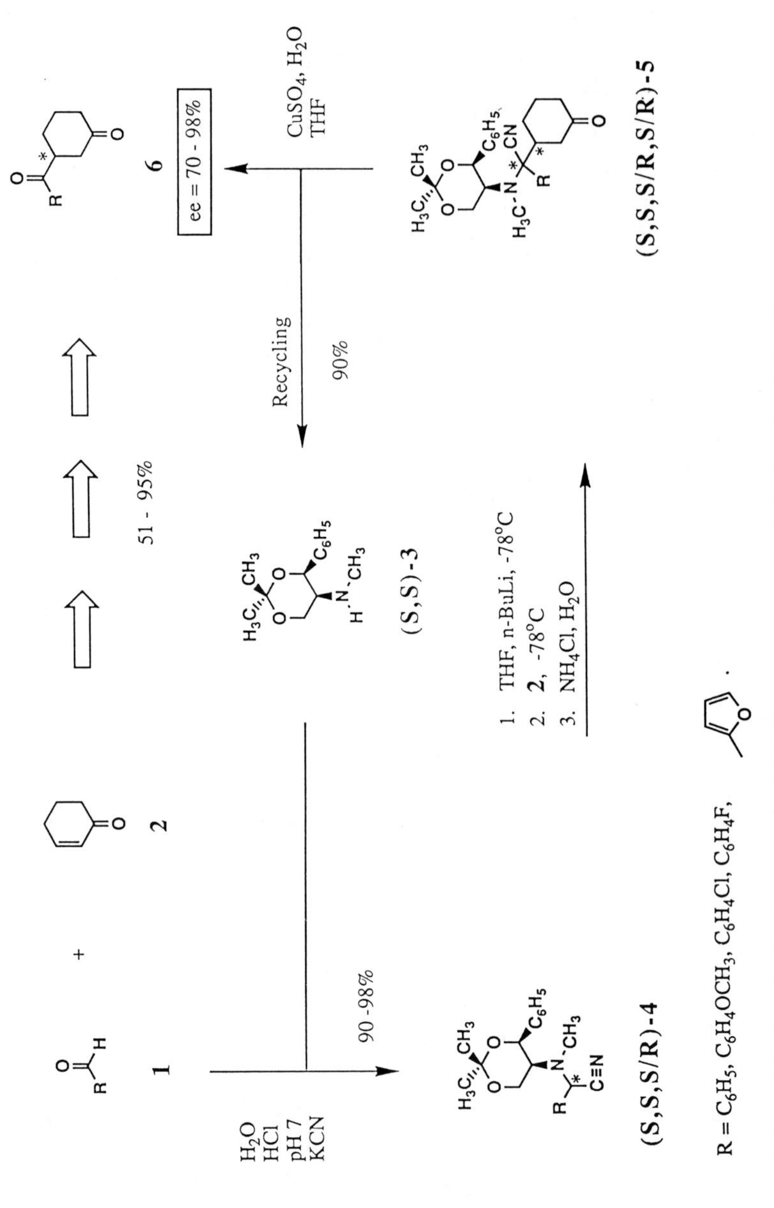

R = C_6H_5, $C_6H_4OCH_3$, C_6H_4Cl, C_6H_4F, .

Scheme 19. Enantioselective synthesis of 1,4-diketones via metalated chiral aminonitriles [?]

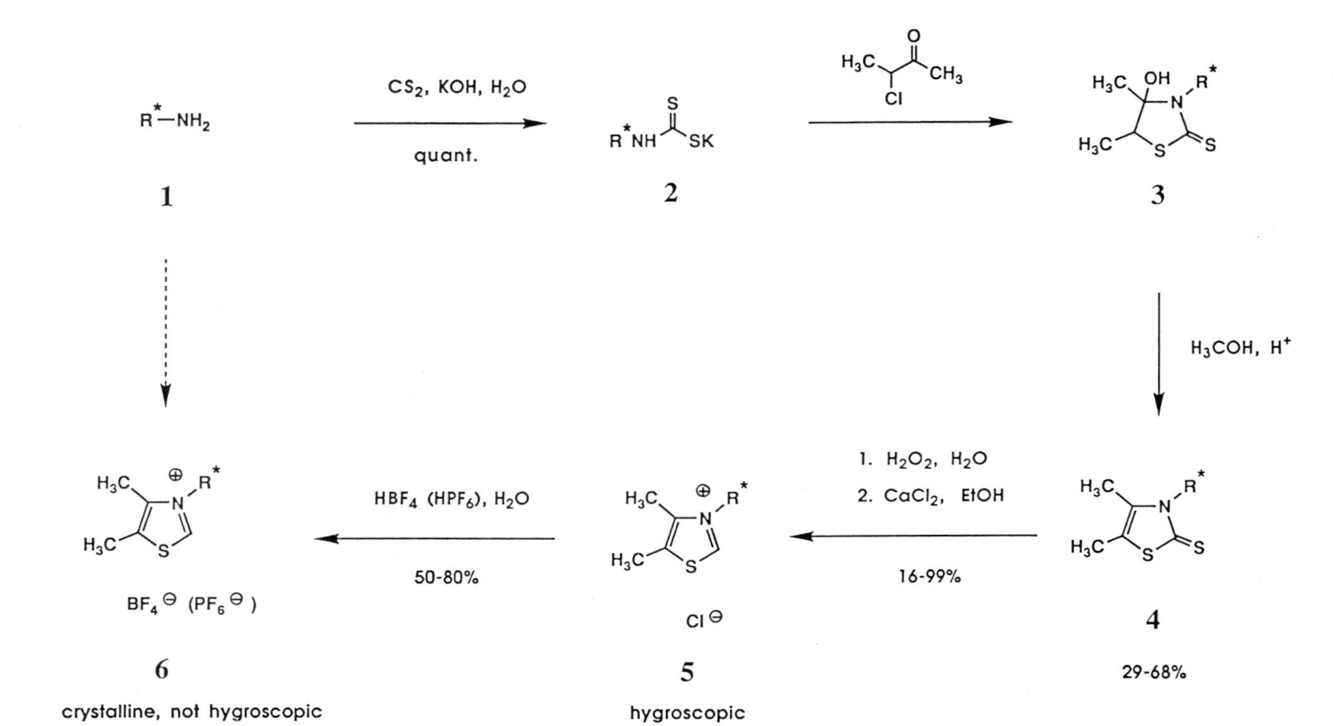

Scheme 20. Efficient and flexible synthesis of chiral thiazolium salts

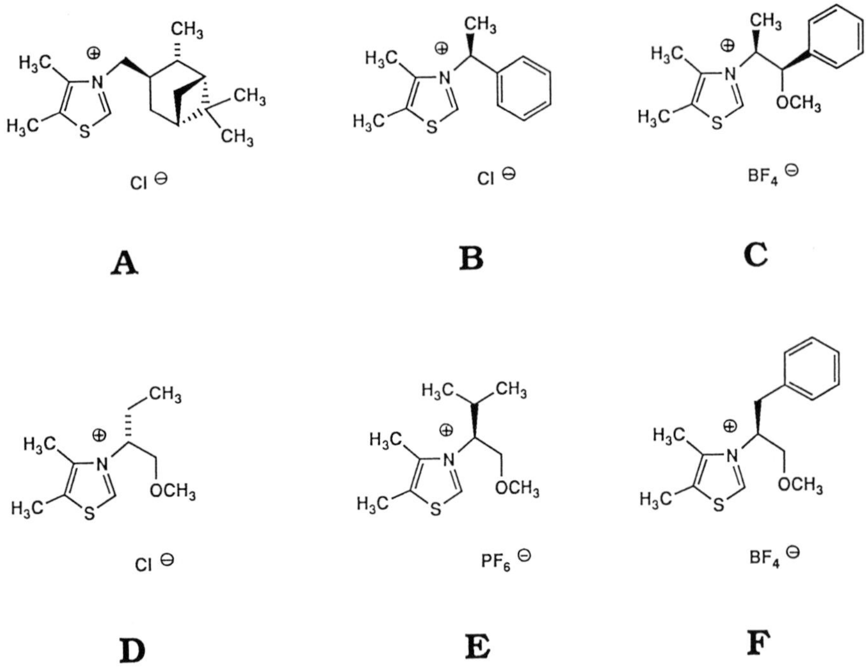

Scheme 21. Novel chiral thiazolium salts - catalysts for enantioselective Stetter reactions and acyloin condensations

which could be handled much easier (scheme 20). Some of the many thiazolium salts we prepared in this manner are shown in scheme 21.

The new enantiopure thiazolium salts were tested in asymmetric Stetter reactions. As was the case in benzoin condensations catalyzed by chiral thiazolium salts and reported many years ago by Sheehan et al. [60,61] and Tagaki et al. [62], only moderate asymmetric inductions were observed. So far the best results are shown in scheme 22, a 30% chemical yield and an enantiomeric excess of 40% [63]. Nevertheless, the principle has been demonstrated, and we are confident to be able to improve the new procedure.

The first chiral synthetic equivalents for the overall enantioselective transfer of the "active glycolaldehyde" moiety were easily prepared from benzyloxyacetaldehyde as chiral aminonitriles (scheme 23) [64], however, the subsequent metalation resulted in β-elimination, so that the active glycolaldehyde mimicry still waits to be solved and will be part of our future projects.

3. Enzymemimetic C-N Bond Formations

Mother Nature is also able to create C-N bonds stereoselectively, for instance, ammonia is added to the double bond of fumaric acid with the help of an aspartase of bacterium cadaveris to form (S)-aspartic acid (scheme 24) [65,66].

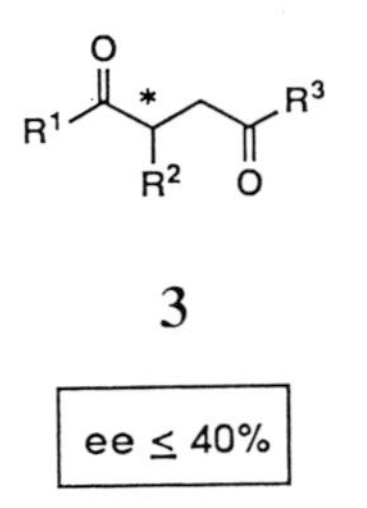

Scheme 22. First test reactions for an asymmetric Stetter reaction via thiazolium salt catalysis [?]

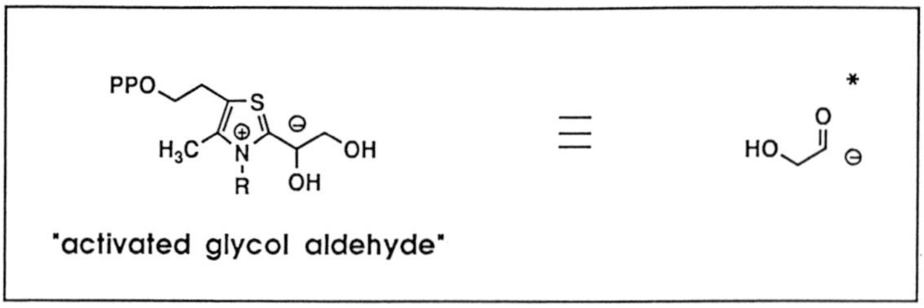

"activated glycol aldehyde"

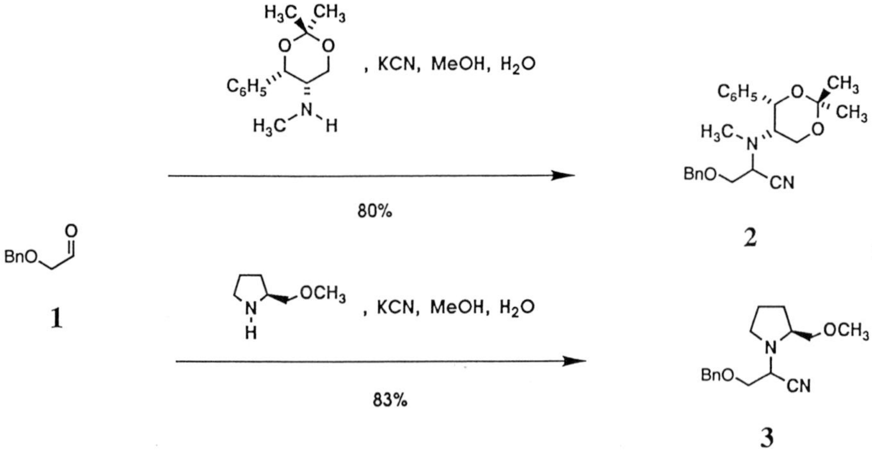

Scheme 23. Chiral equivalents of "active glycolaldehyde"

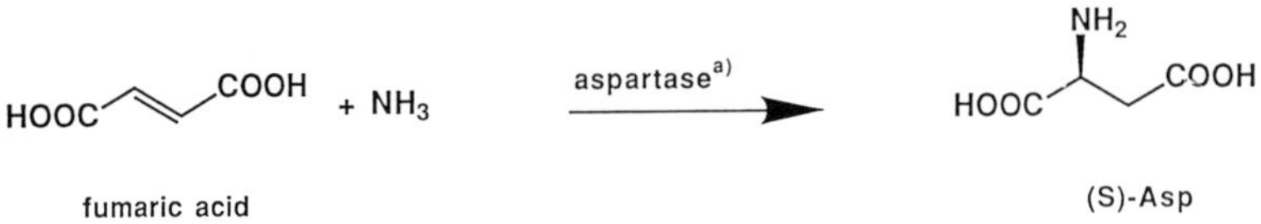

a) From bacterium cadaveris

Scheme 24. Enzymatic synthesis of (S) - aspartic acid

In the eyes of a synthetic chemist this might be called a nitrogen-Michael-addition to the α,β-unsaturated acid moiety as the 1,4-acceptor. Thus, it occurred to us that our favourite auxiliary, the hydrazine SAMP, might be used as a chiral ammonia equivalent in asymmetric hetero Michael additions to enoates.

Indeed, as is shown in scheme 25, N-silylated SAMP (TMS-SAMP) added after lithiation with n-butyllithium in a conjugate fashion to various enoates under highly diastereoselective C-N bond formation. The subsequent N-N bond cleavage by Raney-nickel (recycling of the chiral auxiliary possible) afforded β-amino acids of high enantiomeric purity (ee $\geq$ 90 - 93%) and in good overall yields (30 - 58%) [67]. Most of these β-amino acids form conglomerates and, thus, a single recrystallization gave the enantiomerically pure β-amino acids, im-

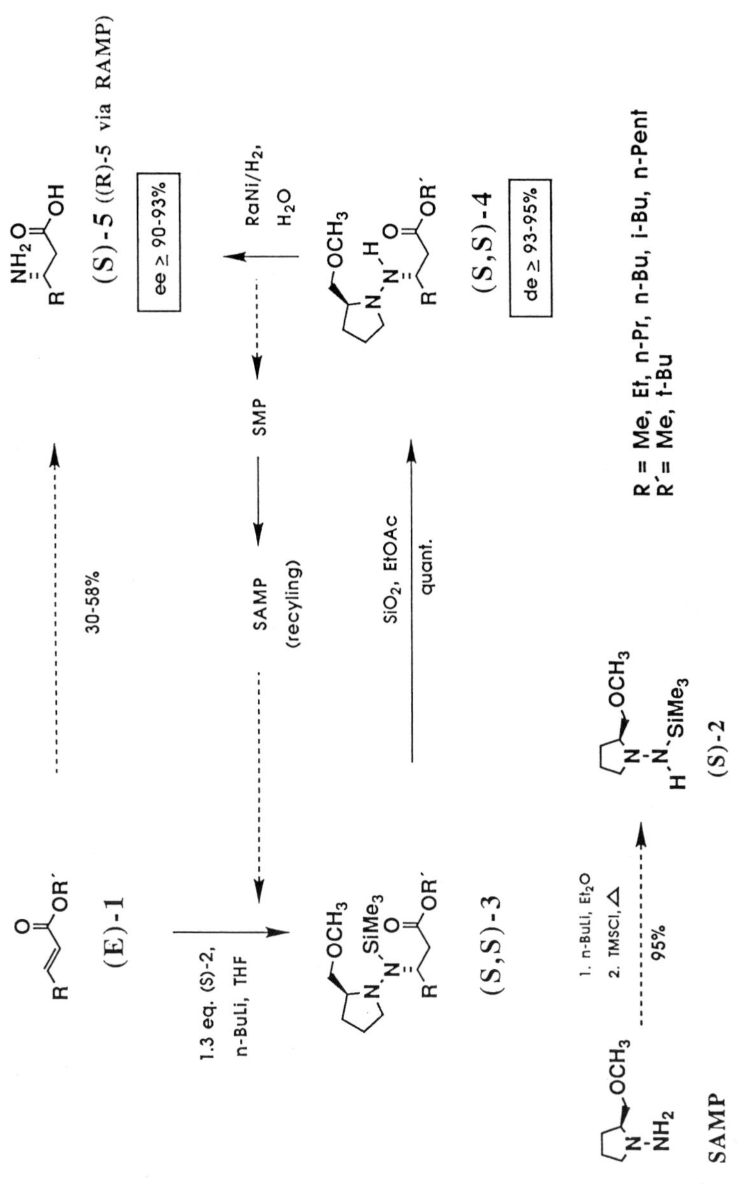

Scheme 25. Efficient enantioselective synthesis of β-amino acids.Asymmetric hetero Michael additions using SAMP as a chiral ammonia equivalent

portant compounds e.g. as precursors of β-lactams. Finally, it should be mentioned at this point that tandem variants of this procedure (N - 1,4/α - alkylation, N - 1,4/aldol reaction) are also possible with full control of up to 3 stereogenic centers [67].

4. Conclusion

This short description of biomimetic C-C and C-N bond formations demonstrates that in a broad range of applications stereoselectivities can be reached, which compare well with the corresponding enzyme reactions. During the early stages of the development of new biologically active compounds, e.g. pesticides, drugs etc., when there is a need to quickly synthesize in a stereochemical flexible way a large number of new structures as both pure enantiomers in gram quantities (screening), the classical chemical synthesis offers advantages. In any given case, future workers must decide which type of synthetic approach they prefer.

Acknowledgement. I would like to thank my able coworkers, whose names are given in the list of references, for their hard work and skill and their enthusiasm to join me in the "chemical mimicry" projects described. I am especially thankful to my colleagues, the Professors Kula, Sahm, Wandrey, Effenberger, Steglich and Winterfeldt, for very stimulating discussions. This work was supported by the Deutsche Forschungsgemeinschaft, the Fonds der Chemischen Industrie and the Land Nordrhein-Westfalen. We thank the chemical industry, Degussa AG, BASF AG, Bayer AG, Hoechst AG, Wacker Chemie, Boehringer Mannheim and Henkel KG, for providing us with chemicals.

References

[1] D. Enders, R.W. Hoffmann, *Chem. Unserer Zeit* **1985**, *19*, 177.
[2] G.M. Whitesides, C.-H. Wong, *Angew. Chem.* **1985**, *97*, 617.
[3] D.G. Drueckhammer, W.J. Hennen, R.L. Pederson, C.F. Barbas, C.M. Gautheron, T. Krach, C.-H. Wong, *Synthesis* **1991**, 499.
[4] E.J. Toone, E.S. Simon, M.D. Bednarski, G.M. Whitesides, *Tetrahedron* **1989**,*45*,5365.
[5] M.D. Bednarski, *Appl. Biocatal.* **1991**, *1*, 87.
[6] M.D. Bednarski, E.S. Simon, N. Bischofberger, W.-D. Fessner, M.-J. Kim, W. Lees,T. Saito, H. Waldmann, G.M. Whitesides, *J. Am. Chem. Soc.* **1989**, *111*, 627.
[7] J.R. Durrwachter, D.G. Drueckhammer, K. Nozaki, H.M. Sweers, C.-H. Wong, *J. Am. Chem. Soc.* **1986**, *108*, 7812.
[8] C.H. von der Osten, A.J. Sinskey, C.F. Barbas, R.L. Pederson, Y.F. Wang, C.-H. Wong, *J. Am. Chem. Soc.* **1989**, *111*, 3924.
[9] T. Ziegler, A. Straub, F. Effenberger, *Angew. Chem.* **1988**, *100*, 737.
[10] A. Straub, F. Effenberger, P. Fischer, *J. Org. Chem.* **1990**, *55*, 3926.
[11] H.-P. Brockamp, M.R. Kula, *Tetrahedron Lett.* **1990**, *31*, 7123.
[12] W.-D. Fessner, G. Sinerius, A. Schneider, M. Dreyer, G.E. Schulz, J. Badia, J. Aguilar, *Angew. Chem.* **1991**, *103*, 596.

[13] D. Enders in Asymmetric Synthesis (Ed.: J.D. Morrison), Academic Press, Orlando, **1984**, vol. 3, p. 275.
[14] D. Enders, B. Bockstiegel, *Synthesis* **1989**, 493.
[15] D. Enders, U. Jegelka, *Tetrahedron Lett.* **1993**, *34*, in press; U. Jegelka, Diploma work, Dissertation, Technical University of Aachen, **1989**, **1992**.
[16] D. Enders, B. Bockstiegel, unpublished results; B. Bockstiegel, Dissertation, Technical University of Aachen, **1989**.
[17] J.A. Dale, H.S. Mosher, *J. Am. Chem. Soc.* **1973**, *95*, 512.
[18] B. Bossow-Berke, Dissertation, University of Bonn, **1989**.
[19] H.B. Kagan, T.P. Dang, J. *Am. Chem. Soc.* **1972**, *94*, 6429.
[20] J.K. Whitesell, *Chem. Rev.* **1989**, *89*, 1581.
[21] D. Enders, W. Gatzweiler, U. Jegelka, *Synthesis* **1991**, 1137.
[22] D. Enders, G. Bachstädter, K.A.M. Kremer, M. Marsch, K. Harms, G. Boche, *Angew. Chem.* **1988**, *100*, 1580.
[23] D. Enders, *Chem. Scripta* **1985**, *25*, 139.
[24] M.G. Koch, AIDS, Vom Molekül zur Pandemie, *Spektrum der Wissenschaften*, Heidelberg, **1989**.
[25] D. Häbich, *Chem. Unserer Zeit* **1991**, *25*, 295.
[26] J.R. Huff, *J. Med. Chem.* **1991**, *34*, 2305; GDCh-Symposium HIV-Infektion, Chemotherapeutische Entwicklungen, Frankfurt am Main, **1992**.
[27] J. Erickson, D.J. Neidhart, J. Van Drie, D.J. Kempf, X.C. Wang, D.W. Norbeck, J.J. Plattner, J.W. Rittenhouse, M. Turon, N. Wideburg, W.E. Kohlbrenner, R. Simmer, R. Helfrich, D.A. Paul, M. Knigge, *Science* **1990**, *249*, 527.
[28] D. Enders, U. Jegelka, B. Dücker, *Angew. Chem.* **1993**, *105*, March issue.
[29] W.D. Busse, lecture at RWTH Aachen, **1992**.
[30] T. Kajimoto, K.K.C. Lin, R.L. Pederson, Z. Zhong, Y. Ichikawa, J.A. Porco, C.-H. Wong, *J. Am. Chem. Soc.* **1991**, *113*, 6187, 6678.
[31] U. Fuhrmann, E. Bause, H. Ploegh, *Biochim. et Biophys. Acta* **1985**, *825*, 95.
[32] S.V. Evans, L.E. Fellows, T.K.M. Shing, G.W.J. Fleet, *Phytochem.* **1985**, *24*, 1953.
[33] G.W.J. Fleet, S.J. Nash, *Tetrahedron Lett.* **1985**, *26*, 3127.
[34] D. Enders, U. Jegelka, *Synlett* **1992**, 999.
[35] D. Enders, D. Kownatka, manuscript in preparation; D. Kownatka, Dissertation, Technical University of Aachen, **1992**.
[36] F.M. Unger, *Adv. Carbohydr. Chem. Biochem.* **1980**, *38*, 323.
[37] R. Schauer, *Adv. Carbohydr. Chem. Biochem.* **1982**, *40*, 132.
[38] R. Schauer, *Pure Appl. Chem.* **1984**, *56*, 902.
[39] S.T. Allen, G.R. Heintzelmann, E.J. Toone, *J. Org. Chem.* **1992**, *57*, 426 and literature cited therein.
[40] U. Kragl, D. Gygax, O. Ghisalba, C. Wandrey, *Angew. Chem.* **1991**, *103*, 854.
[41] D. Enders, H. Dyker, G. Raabe, *Angew. Chem.* **1992**, *104*, 649.
[42] D. Enders, H. Dyker, G. Raabe, J. Runsink, *Synlett* **1992**, 901.
[43] J.E. Hengeveld, V. Grief, J. Tadenier, C.M. Lee, D. Riley, P.A. Lartey, *Tetrahedron Lett.* **1984**, *25*, 4075.
[44] V. Faivre, C. Lila, A. Saroli, A. Doutheau, *Tetrahedron* **1989**, *45*, 7765.
[45] D. Enders, H. Dyker, G. Raabe, *Angew. Chem.* **1993**, *105*, March issue.
[46] D. Enders, H. Kipphardt, P. Gerdes, L.J. Brena - Valle, V. Bhushan, *Bull. Soc. Chim. Belg.* **1988**, *97*, 691.
[47] H. Eichenauer, E. Friedrich, W. Lutz, D. Enders, *Angew. Chem.* **1978**, *90*, 219.
[48] D. Enders, H. Eichenauer, R. Pieter, *Chem. Ber.* **1979**, *112*, 3703.
[49] M. Braun, *Angew. Chem.* **1987**, *99*, 24.
[50] D. Enders,U. Baus, K.A.M. Kremer, University of Bonn, unpublished results; U. Baus, Dissertation, University of Bonn, **1985**.
[51] D. Enders, B.B. Lohray, *Angew. Chem.* **1988**, *100*, 594.
[52] H. Lotter, Diploma work, University of Giessen, **1979**; Dissertation, University of Bonn, **1985**.
[53] D. Enders, H. Lotter, *Angew. Chem.* **1981**, *93*, 831.

[54] D. Enders in *Current Trends in Organic Synthesis* (Ed.: H. Nozaki), Pergamon Press, Oxford, **1983**, p. 151.
[55] D. Enders, H. Lotter, N. Maigrot, J.-P. Mazaleyrat, Z. Welvart, *Nouv. J. Chim.* **1984**, *8*, 747.
[56] D. Enders, P. Gerdes, H. Kipphardt, *Angew. Chem.* **1990**, *102*, 226.
[57] D. Enders, D. Mannes, G. Raabe, *Synlett* **1992**, 837.
[58] D. Enders, P. Gerdes, unpublished results; P. Gerdes, Dissertation, Technical University of Aachen, **1989**.
[59] W. Tagaki, Y. Tamura, Y. Yano, *Bull. Chem. Soc. Jpn.* **1980**, *53*, 740.
[60] J. Sheehan, D.H. Hunneman, *J. Am. Chem. Soc.* **1966**, *88*, 3666.
[61] J. Sheehan, T. Hara, *J. Org. Chem.* **1974**, *39*, 1196.
[62] W. Tagaki, Y. Tamura, Y. Yano, *Bull. Chem. Soc. Jpn.* **1980**, *53*, 478.
[63] D. Enders, K. Papadopoulos, H. Kuhlmann, J. Tiebes, unpublished results.
[64] D. Kownatka, Dissertation, Technical University of Aachen, **1992**.
[65] V.R. Williams, R.T. McIntyre, *J. Biol. Chem.* **1955**, *217*, 467.
[66] S. Englard, *J. Biol. Chem.* **1958**, *233*, 1003.
[67] D. Enders, H. Wahl, W. Bettray, unpublished results; H. Wahl, Dissertation, Technical University of Aachen, **1990**; W. Bettray, Dissertation, Technical University of Aachen, **1993**.

Synthesis of Natural Products of Polyketide Origin, An Exemplary Case

R. W. HOFFMANN AND R. STÜRMER

Fachbereich Chemie der Philipps-Universität,
Hans-Meerwein-Strasse, D-35043 Marburg/Lahn

Summary: A short linear synthesis of erythronolide is discussed as an exemplary case to demonstrate how changes in the attitude towards new synthetic methods increased the efficiency of natural product synthesis.

1. Aims of Natural Product Synthesis, Changes over Time

The synthesis of natural products of ever increasing complexity has been a perennial task of preparative organic chemistry. However, over time, the motivation underlying this activity has changed in a characteristic manner. Up to the 1960s the synthesis of a natural product constituted the sole - and later on, the most important - proof of its structure. The reliability of such structural proof was not 100% - as can be seen from the structural assignment of patchouli-alcohol **1**, a case in which the wrong structure has been "proven" by synthesis [1]. The correct structure was finally established by X-ray analysis [2].

As long as the aim of natural product synthesis was to prove the proposed constitution and configuration of a new natural product, it was mandatory to use only synthetic transformations which were fully established in terms of their scope and reliability. The result could otherwise have been considered as ambiguous. As a consequence, natural product synthesis had to rely on a rather restricted arsenal of established methods. The task of providing structural proof was taken over gradually in the period from 1930 to 1970 by X-ray crystal structure analysis and NMR-spectroscopy. At the same time natural product synthesis was freed from the confinement regarding the methods to be used [3]. There-

OH
1

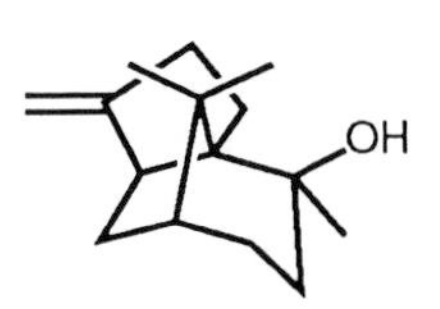

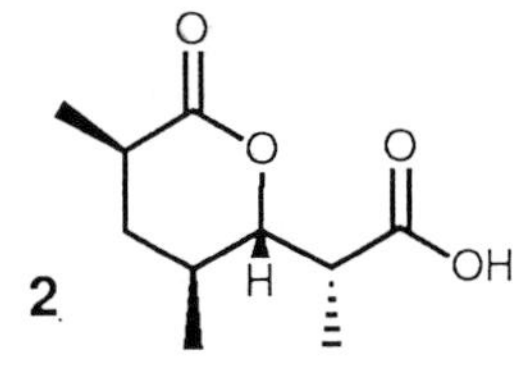

Patchouli alcohol — Structure established by X-ray - analysis

"Patchouli alcohol" — proposed structure proven by synthesis

Prelog - Djerassi - lactonic acid

Scheme 1

Stereoselective Synthesis
Editors: Ottow, Schöllkopf, Schulz

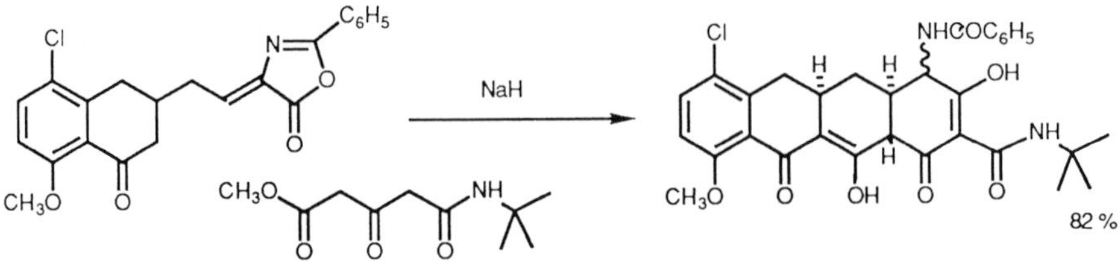

Scheme 2

after, novel synthetic methods and imaginative synthetic concepts could be applied. The first "modern" synthesis of this kind is in my opinion the synthesis of the tetracycline skeleton by Muxfeldt [4]. Muxfeldt used an unprecedented transformation in which three C-C-connections were generated in a single operation. Such a rapid increase in complexity [5] is the hallmark of an efficient synthesis.

The new situation in natural product synthesis in the seventies led to a burst of development in new synthetic methods and in natural product synthesis as such. When a multitude of related synthetic methods were available for application in natural product synthesis, it was not a question whether a natural product could be synthesized, but rather how efficiently such a synthesis could be carried out. Efficiency was roughly taken as the number of steps needed to reach a certain target. Among these steps, those which construct the molecular backbone of the target are essential. All other steps such as refunctionalizations and introduction or removal of protective groups merely reflect our inability to carry out syntheses with ideal efficiency. An optimum synthesis [6] is one in which the only steps involved are those that assemble the molecular skeleton. These steps have to be such, that they directly create the functionality required in the final structure or are necessary to carry out the next step. Likewise, all the stereocenters have to be set up with the correct configuraton in the structure building steps. The efficiency of a synthesis could be further increased, if not only a single, but e.g. three or four stereocenters are generated in one step; if not only a single carbon-carbon bond, but three or four are made in a single operation. This is why cycloaddition reactions [7] or tandem processes [8] have become so important for synthetic efficiency. In this quest for efficiency in synthesis, high demands are put on the stereoselectivity and chemoselectivity of the methods to be developed. This has been perfectly articulated in the title of a symposium held in the eighties: "Selectivity - a Goal for Synthetic Efficiency" [9]. Efficiency in synthesis can be measured by various abstract parameters [5]. It is, however, much more illustrative and convincing to compare different syntheses of the same natural product. This is the reason why an increasing number of syntheses have been directed at a single natural product or a single synthetic intermediate (for example the Prelog-Djerassi-lactone **2** [10]) even if there was no actual need for such a synthesis.

Natural product synthesis in the eighties thus became the testing ground for newly developed synthetic strategies and methods. In this manner, the situation in natural product synthesis became more and more comparable to that in mountaineering in which, as both an artistic and technical discipline, the outfitters (those that developed the methods) contribute to the success and efficiency. Simply because natural product synthesis is unimportant to the media, there is as yet no commercial advertising on the T-shirts of those practising natural product synthesis. What for mountaineers at one time was the north slope of the Eiger has been in natural product synthesis the efforts devoted towards erythronolides. The dimension of this challenge has been accentuated by the early statement of Woodward in 1956 [11]: "Erythromycin, with all our advantages, looks at present hopelessly complex, particularly in view of its plethora of asymmetric centers." The first successful synthesis of erythronolide B and erythronolide A by Corey [12] and by Woodward [13] were scientific achievements comparable to the first ascent of a Himalayan mountain. These landmark syntheses were followed by an apparently never ending sequence [14] of subsequent efforts. These were driven by the aim to develop specific methods for the synthesis of such natural products of polyketide origin, hopefully allowing more and more efficient syntheses. This holds equally true for our synthesis of 9(S)-dihydroerythronolide A, [15] which was recently completed.

2. New Methods for the Synthesis of Polyketide Derived Natural Products

Erythronolide A **3** is a derivative of a C_{15} fatty acid [16] which carries at every second carbon a methyl branch constituting a stereogenic center. Oxygen functionalities are found at most of the even numbered carbon atoms leading to stereocenters at C3, C5, C11, and C13. In addition there are two tertiary stereogenic centers at C6 and C12.

An efficient synthesis of such structures can be envisioned by using an iterative approach. In analogy to the biosynthesis [17], propionyl units could be linked one after another. However, in order to attain a 1,3,5,n-polyhydroxylated chain, an aldol-addition appears to be the method of choice [18]. The starting point is an aldehyde **4**. Addition of an enolate to this aldehyde results in chain extension and generates both the alcohol function and the methyl branching.

This method entails two subsequent non-productive steps: the protection of the hydroxyl group in **5** and a refunctionalization which generates a new alde-

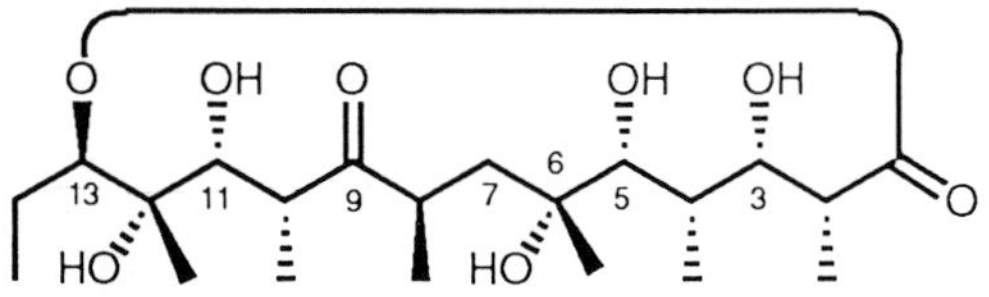

O OH O OH OH
13 11 9 7 6 5 3
HO HO
Erythronolide A
3

Scheme 3

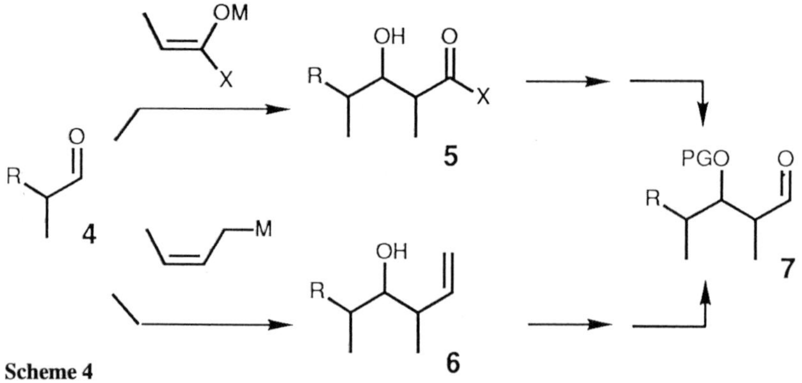

Scheme 4

hyde function. The resulting aldehyde **7** can then be used as the starting point for another similar chain extension. In keeping with the demands of efficiency, both stereocenters are generated in the steps that form the carbon skeleton. This of course requires perfect stereocontrol in this particular step. Remarkable progress has been made in this direction as is demonstrated by the syntheses carried out by Masamune [19], Heathcock [20] and Paterson [21] in the erythronolide series.

We turned to a study of an analogous transformation, the crotylmetallation of aldehydes, in which the homoallyl alcohol **6** is formed from **4**. Again two steps that decrease the synthetic efficiency follow in order to reach the homologous aldehyde **7**. The main demand is placed again on stereocontrol: A total of 4 stereotriads A-D [22] may result on reaction of an α-methylbranched aldehyde **8** with a crotylmetallic species. Therefore such chain extensions will be attractive only if the method allows for conversion of a given aldehyde **8** selectively into each of the four stereotriads A - D at will.

This is exactly what can be achieved by the crotylboration reaction using chiral crotylboronates. Crotylboronates add to aldehydes displaying high simple diastereoselectivity, i.e. E-crotylboronates **9** lead to the *anti*-homoallylic alcohols **10**, whereas Z-crotylboronates **11** generate the *syn*-homoallylic alcohols **12** [23].

We planned an iterative synthesis of erythronolide A (**3**), a synthesis in which the molecule grows from the left end (C15) to the right end (C1). This takes into account the fact that only syn-β-methyl-alcohol units have to be generated which correspond to the stereotriads A and B. These however, have to be generated with the correct absolute configuration. One has to reckon that the

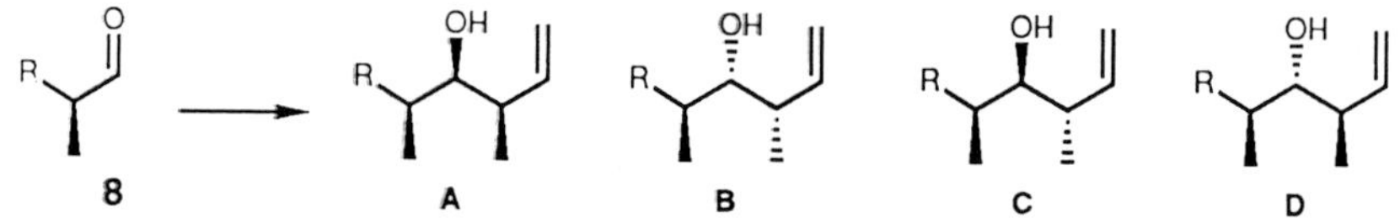

Scheme 5

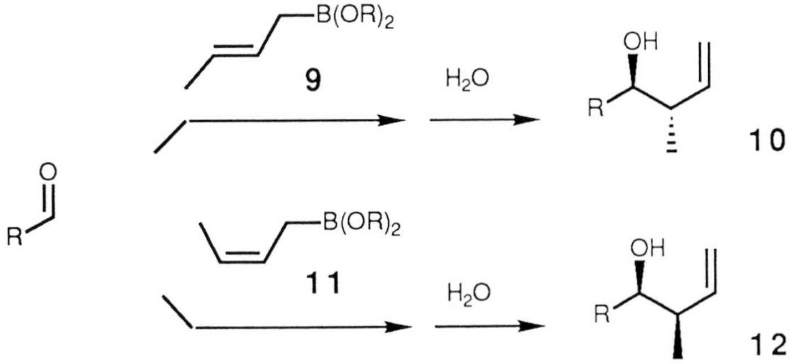

Scheme 6

starting aldehyde is chiral. Hence, asymmetric induction originates from this chiral substrate favoring the formation of the stereotriad B with moderate selectivity over the stereotriad A. An increased selectivity in favor of B and - more difficult - a reversal in selectivity in favor of A will only be possible by application of external asymmetric induction based on a chiral reagent. In fact, the asymmetric induction of the reagent has to be so pronounced, that it overrides the asymmetric induction of the aldehyde (the substrate) to achieve what is called "reagent control of diastereoselectivity" [24]. For this purpose we developed α-substituted chiral crotylboronates such as the pentenylboronate **13** [25]. Asymmetric induction on reaction of such a crotylboronate with an aldehyde profits from the fact that the reaction proceeds via a cyclic six-membered transition state. Two transition states, **14** and **16**, have to be considered for reaction of a given enantiomer of the reagent **13** with the two enantiotopic faces of the aldehyde. These two transition states lead correspondingly to products **15** and **17** of opposite absolute configuration at the two new stereocenters.

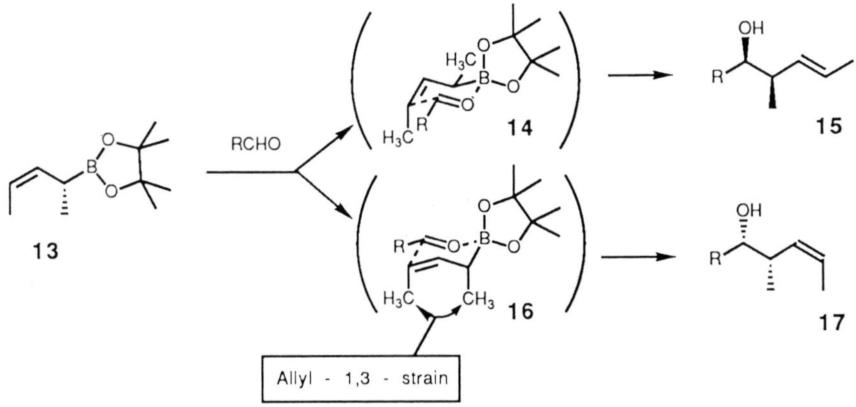

Allyl - 1,3 - strain

Scheme 7

Asymmetric induction results, as soon as the two transition states **14** and **16** are of unequal energy. The transition states differ in the arrangement of the substituent in the α-position of the reagent: In **14** the methyl group takes an equatorial position, while in **16** it is in an axial position. The axial methyl group eclipses the Z-positioned methyl group at the double bond. This causes a destabilization by allylic 1,3-strain of ca. 3.5 kcal [26] and, as a consequence, the reaction proceeds exclusively - and that is born out by the experiment - via transition state **14**.

Application of these pentenylboronates in asymmetric synthesis requires enantiomerically pure reagents. In fact, both enantiomeric forms are required for our planned synthesis of erythronolide A. Both enantiomers of **18** have become available by use of the easily accessible dicyclohexylethanediol as chiral auxiliary.

As expected, reagent **18** added to a series of aldehydes with excellent asymmetric induction [25].

Coming back to scheme 7 one notes that reaction via transition state **14** leads to a homoallylic alcohol **15** with an E-double bond. In turn, reaction via transition state **16** generates a product **17** having a Z-double bond. The geometry of the double bond and the configuration of the newly formed stereocenters are therefore mechanistically linked. The geometry of a double bond is much more readily determined than the relative or absolute configuration of an alcohol. The geometry of the newly formed double bond may thus be used to advantage as an indicator for the configuration of the new stereocenters.

The stereoselective crotylboration reaction allows the easy generation of homoallylic *alcohols*. When considering the synthesis of erythronolide A it is therefore appropriate to envision 9-dihydroerythronolide A (**19**) as a late intermediate, especially since its conversion into erythronolide A (**3**) is known [27]. When a Z-pentenylboronate reagent is used to generate the methyl bearing stereocenter at C8, the intermediate reached will have the (S)-configuration at C9. Thus the 9(S)-dihydroerythronolide A **19**, in which the appendages on C8 and C9 are syn, was the primary target of our synthesis.

$Cl_2CH-B(O_2C_2H_2(C_6H_{11})_2)$ $\xrightarrow[ZnCl_2]{CH_3Li}$ $CH_3CH(Cl)-B(O_2C_2H_2(C_6H_{11})_2)$ $\xrightarrow{\text{(Z)-pentenyl-Li}}$ **18**

Scheme 8

18 $\xrightarrow{RCHO}$ homoallylic alcohol (R–CH(OH)–CH(CH_3)–CH=CH–CH_3)

R		Yield	e.e.
=	C_6H_5-	71 %	99.5 % e.e.
=	CH_2=C(CH_3)-	86 %	99.0 % e.e.
=	CH_3-CH_2-	79 %	98.5 % e.e.

Scheme 9

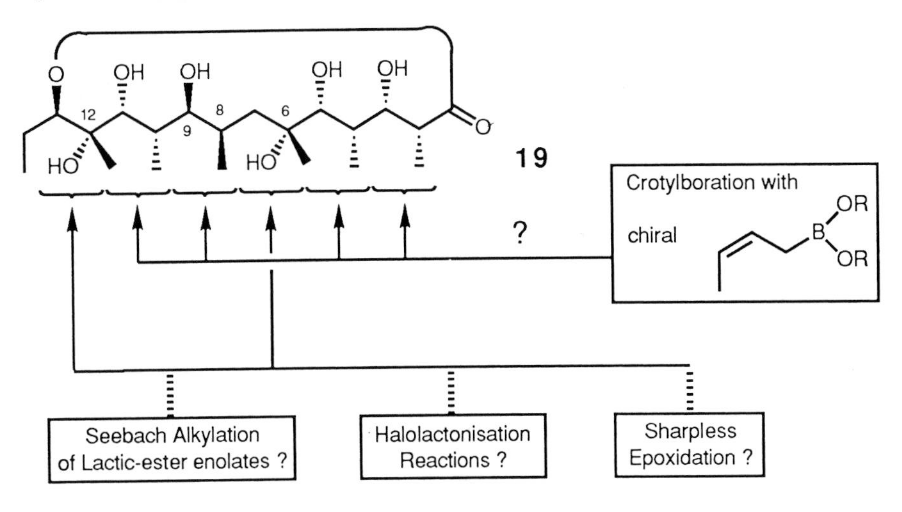

Scheme 10

Scheme 11

A quick analysis of 9(S)-dihydroerythronolide A **19** shows that 7 of the 11 stereocenters could be created using reagent control of diastereoselectivity by crotylboration reactions. The stereogenic centers at C6 and C12 having a tertiary alcohol function are presently outside the scope of stereoselective crotylboration reactions. Here we have to rely on other methods. For instance, the use of lactic acid enolates developed by Seebach [28] appeared attractive to generate the tertiary centers both at C6 and C12.

Unfortunately, hydroxyalkylation of the enolate **20** by propionaldehyde led to a 1:1 diastereomeric mixture of **21**. This entailed two additional steps (oxidation and stereoselective reduction; ds 90%) to reach diastereomerically pure material **22**.

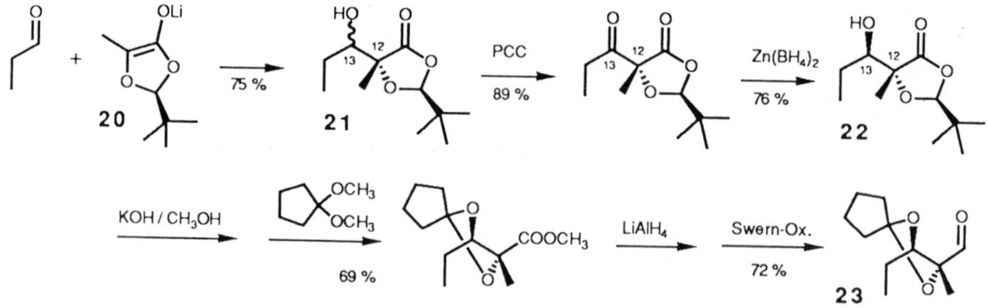

Scheme 12

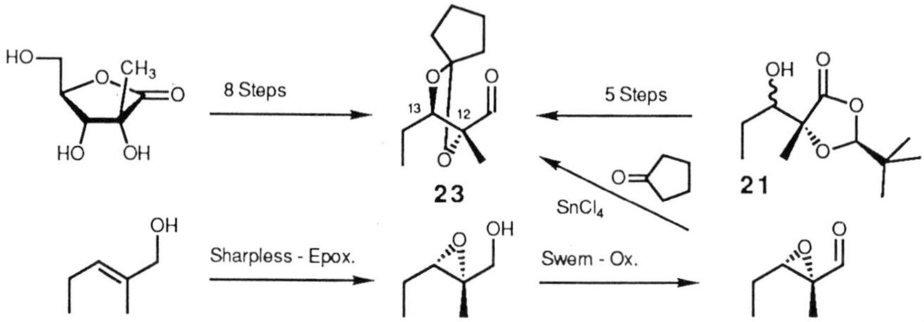

Scheme 13

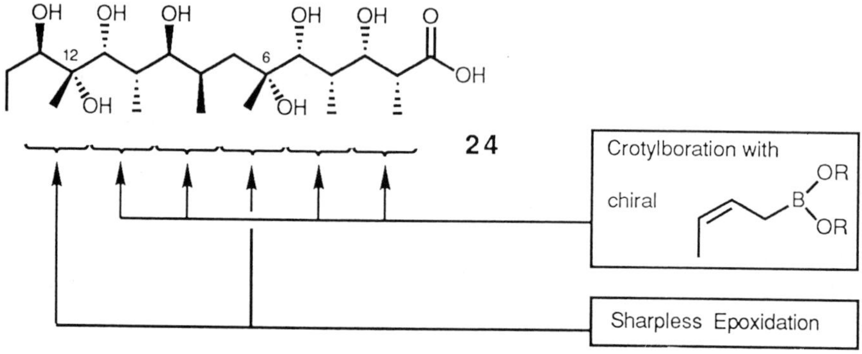

Scheme 14

The synthesis of the key starting material **23** via this route (involving up to 8 steps) was therefore as inefficient as our earlier way to **23** starting from fructose [29]. The expenditure of 8 steps for a building block with just two stereocenters could clearly not meet the demands of an efficient synthesis. For this reason we explored an access to **23** via the Sharpless epoxidation. This resulted in a concise three step synthesis of **23**.

The Sharpless epoxidation turned out to be similarly successful in the generation of the tertiary stereocenter at C6 when compared to other routes such as the Seebach-alkylation of lactic acid enolates or iodolactonization procedures.

The methods described above should allow an efficient assembly of the molecular skeleton of erythronolide A. This would involve six chain elongation steps leading to the seco-acid **24** of 9(S)-dihydroerythronolide A.
We realized that a linear synthesis is generally less efficient than a convergent one. Therefore a number of convergent variants of the above synthetic plan were considered and tested. These efforts were discontinued when it was found that these convergent approaches to erythronolide could theoretically be only marginally more (and in practice less) efficient than the linear synthesis described below.

3. Linear Synthesis of a Protected seco-Acid of 9(S)-Dihydroerythronolide A

Application of the above mentioned methods led to a concise synthesis of the seco acid of 9(S)-dihydroerythronolide A **24** [15] as is shown in scheme 15.

The crotylboration reaction proved reliable for the generation of the stereocenters at C2, C3, C4, C5, C8, C9, C10, and C11. The high stereoselectivity in the conversion of **23** into **25** is remarkable in view of the inverse stereoselectivity for aldol addition to aldehydes of this type **23** [30]. It is apparent that our chiral reagent **18** was in a position to override the asymmetric induction originating from the aldehyde **23**. In the next chain extension, the enantiomeric reagent ent-**18** served perfectly to place the hydroxyl group and methyl group on the β-side of the chain molecule. The next chain extension serves to furnish the prochiral group for the generation of the tertiary center at C6. We therefore used a Wittig-reaction of the aldehyde derived by ozonolysis of **27**, followed by reduction to generate the allylic alcohol **28**. After Sharpless epoxidation of **28** and oxidation to the aldehyde, chain extension with the pentenylboronate **18** proceeded smoothly. It is the next chain extension **31** - **32** which must work against the high asymmetric induction of the aldehyde **31**. This aldehyde prefers the formation of the Cram-product (not shown) while the anti-Cram-product **32** has to be obtained. The steric situation in the approaches to the two diastereotopic faces of the aldehyde are shown in the following scheme. Attack of the crotyl moiety via transition state **34** to generate the anti-Cram-product **35** occurs alongside the bulky C5/C6-unit. A much more favorable approach from the view point of the aldehyde is via transition state **36** in which the boronate approaches the aldehyde alongside the C4-methyl group. However, this requires the boronate to take up a conformation which suffers from serious allylic 1,3-strain.

The reaction led to two products. The major one had an E-configurated double bond from which we conclude that it must be the anti-Cram isomer **35**. Therefore reagent control of diastereoselectivity dominated the reaction. The minor product of the reaction contained a Z-double bond. It must arise via transition state **36** and should therefore be the product of substrate control of diastereoselectivity.

OH

3 steps → 23 (69 %)

18: 3d, benzene, 80 °C → 4 (81 %, ds. > 95 %)

pMBCl, NaH, DMF (97 %)

a) O_3, -78 °C b) Ph_3P, CH_2Cl_2

pMBO

ent-18: 2d, petr.ether, 25 °C → 26 (79 %)

DDQ, mol.sieves, 0 °C → 27 (91 %)

pMPh

a) O_3, -78 °C b) Ph_3P, CH_2Cl_2

$Ph_3P^+CH^-(CH_3)COOEt$, 1d, CH_2Cl_2

$LiAlH_4$, Et_2O (91 %)

28 (99 %)

tBuOOH, $Ti(OiPr)_4$, 2 eq. (+) dimethyl tartrate, CH_2Cl_2,-30 °C → 29 (90 %, *d.r.* 8 : 1)

NMO, cat. $nPr_4N^+RuO_4^-$ (96 %)

18: 3d, 10 kbar, petr.ether → 30 (79 %)

a) iPrMgCl, b) 5 eq. $LiAlH_4$, 6h THF 65 °C (97 %)

pMBCl, NaH, DMF (72 %)

a) NMO, cat. OsO_4 b) $NaIO_4$ → 31

18: 3d, 10 kbar, petr.ether → 32 (70 % (along with another diastereomer))

DDQ, mol.sieves, CH_2Cl_2, -20 °C → 33 (82 %)

Cy = cyclohexyl; **pMB** = *p*-methoxybenzyl; **pMPh** = *p*-methoxyphenyl;
DDQ = dichloro-dicyano-*p*-quinone; **NMO** = *N*-methyl-morpholine-*N*-oxide.

Scheme 15

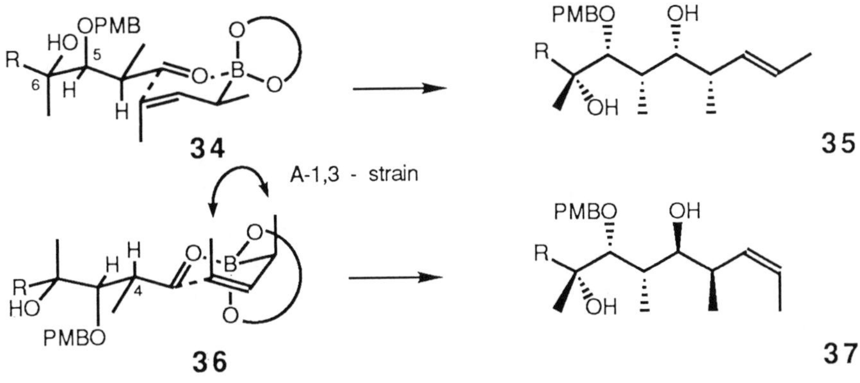

Scheme 16

All in all we successfully generated the 11 stereocenters of 9(S)- dihydroerythronolide A in a 20 step sequence. The stereoselective chain extension reactions were not the ones which caused difficulties. Difficulties were encountered at the refunctionalization steps: We had to try very hard to realize the reductive opening of the C6-C7-epoxide, a transformation in which a whole set of standard reagents failed, both at the stage of **29**, **30** or its O-protected derivatives. Finally, internal activation via a magnesium chelate allowed for the long sought after transformation of **30** into **31**.

4. The Macrolactonization to 9(S)-Dihydroerythronolide A

The lactonization of erythronolide A seco acid to form a 14 membered lactone has a cumbersome and instructive history. The Woodward group tested a plethora of protecting group patterns at the seco acid stage until cyclization was finally realized [13] with **38**. The success has been traced to the 9(S) configuration of the cyclization precursor **38**.

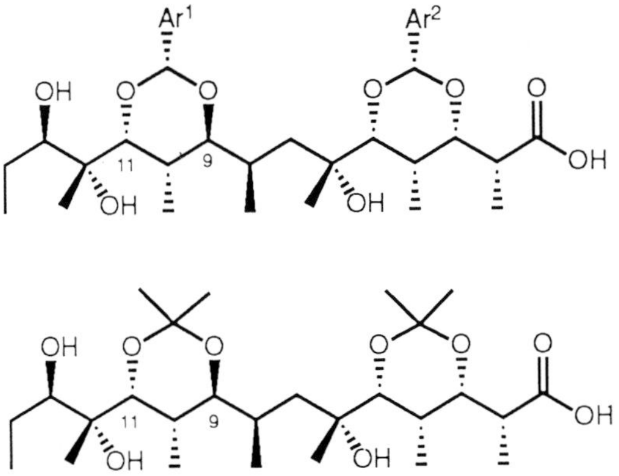

38: Ar¹ = Ar² = mesityl (Woodward)

39: Ar¹ = mesityl, Ar² = p-methoxyphenyl (Yonemitsu)

40: (Stork)

Scheme 17

It is a typical property of polypropionate derived natural products that the methyl branches along the main chain restrict the conformational freedom of the main chain [31]. The few precursor conformations populated may not be the ones which allow a facile macrolactonization. Once the conformation required for macrolactonization is not an energetically favored one, even application of the dilution principle may not coax the ring closure to proceed. This is illustrated with the seco acid **40** which, if at all [32], cyclized only with minimal yield [33]. Calculations [34] show that **40** populates conformation **40A**, which is not on the cyclization pathway. An optimal conformation for the cyclization of an erythronolide seco acid can be realized with the 9(S)-dihydro-derivatives [13] if C9 and C11 are bridged by a dioxane ring with a chair conformation. This causes C8 to occupy an axial position [32] cf. **41**. The unimportant looking difference between **38** and **40** is that the trans-disubstituted C9/C11-acetonide takes up a twist boat conformation [35] and this precludes the necessary bend of the main chain to facilitate lactonization. The axial arrangement of C8 on the C9/C11 dioxane ring not only causes a bending of the main chain, but also forces C7 to be arranged gauche and not anti to C10, cf. **41**. This is the consequence of having a branched axial substituent on a cyclohexane ring [31].

Likewise, a single preferred conformation of the C2/C3-bond is caused by the C3/C5 dioxane ring, cf. **44**. Here, the axial methyl substituent at C4 induces a gauche arrangement of C1 relative to C4 in order to avoid unfavorable g[+[g]-] interactions. This orientation of the carboxyl group brings it within binding distance for macrolactonization.

How does one then attain an axial orientation of C8 at C9/C11 dioxane ring? This requires the aryl-residue Ar in **41** to be placed in an equatorial position, which is unfortunately the thermodynamically less stable situation. As soon as equilibration becomes possible, **41** is converted into **42**, which goes over into the twist boat conformation **43** [36] in which all the major residues are equatorially arranged. Erythronolide seco acid derivatives with such an arrangement have a

41 **42** **43**

Ar = *p*-methoxyphenyl

44

Scheme 18

minimal tendency towards lactonization, whereas those having the configuration shown in **41** can readily be cyclized [36]. In fact, calculations demonstrate that in **39** the conformation leading to cyclization [34] is already populated to 3%.

The dioxane ring with the substitution pattern as in **41** has to be generated under kinetic control. Woodward, [13] as well as Yonemitsu [36], used a transacetalization of the C9/C11-diol with the acetal of 2,4,6-trimethylbenzaldehyde. Stork [32] in his synthesis of 9(S)-dihydroerythronolide A used a stereoselective reduction of an ortho-acetate **47**; in which the hydride was delivered from the

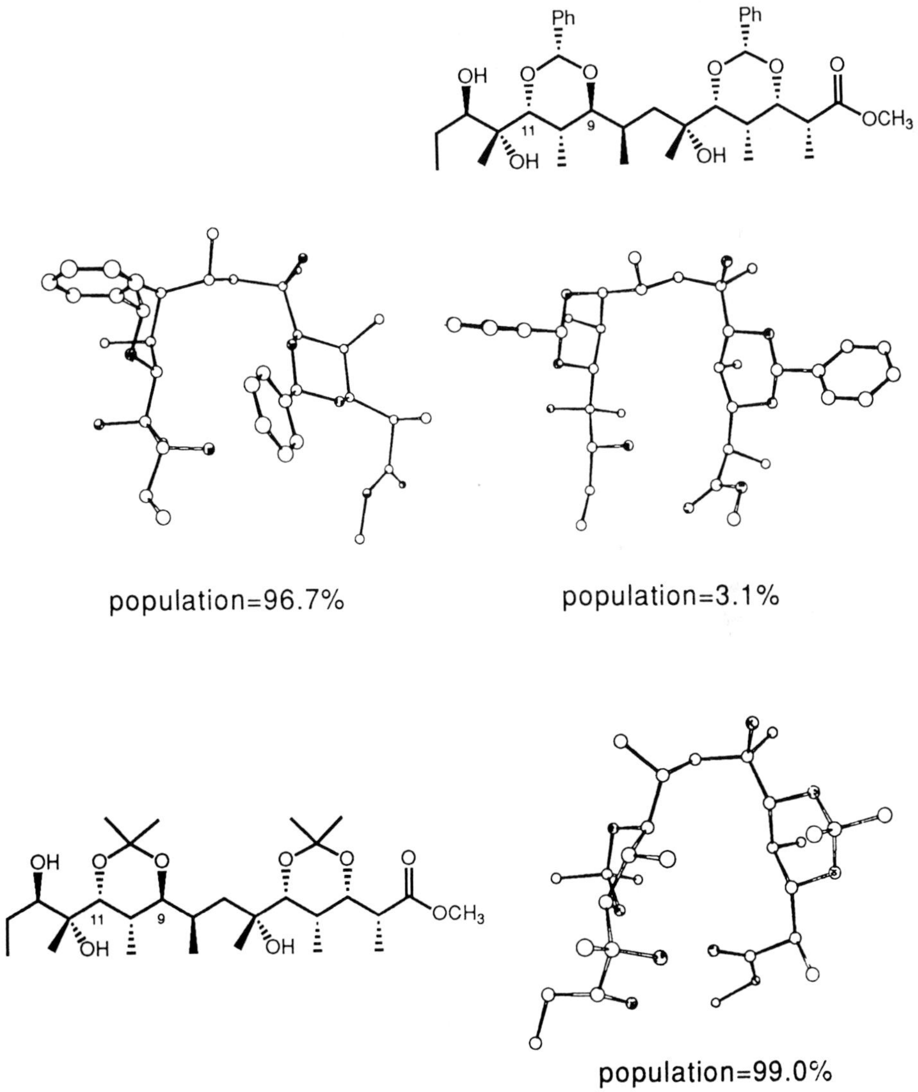

population=96.7%

population=3.1%

population=99.0%

Scheme 19 (We thank Prof. O. Yonemitsu for the diagrams.)

axial direction. We hoped to reach structure **41** by a kinetically controlled acetalization under non acidic conditions.

It is known [37] that the DDQ-oxidation of a p-methoxybenzyl ether **45** generates an oxonium ion **46** in its low energy conformation in which the two hydrogen atoms are in a syn disposition. This conformation of the oxonium ion avoids 1,3-allylic strain [26]. Cyclization via this conformation under non equilibrated conditions leads directly to the desired acetal **41**. Once **41** was formed, we had to avoid acidic conditions during the rest of the synthesis. We were therefore dismayed when the dichloro-dicyano-hydroquinone formed as a coproduct on DDQ-oxidation of the acetal **32** led to an epimerization of the C9/C11-dioxane! The resulting product **48** cannot be recycled and is completely worthless for the synthesis of erythronolide.

This mishap could eventually be avoided by using a specially developed reagent, DDQ on molecular sieves, which allowed the transformation of **32** into **33** without epimerization of the strained [38] and labile dioxane ring. Neverthe-

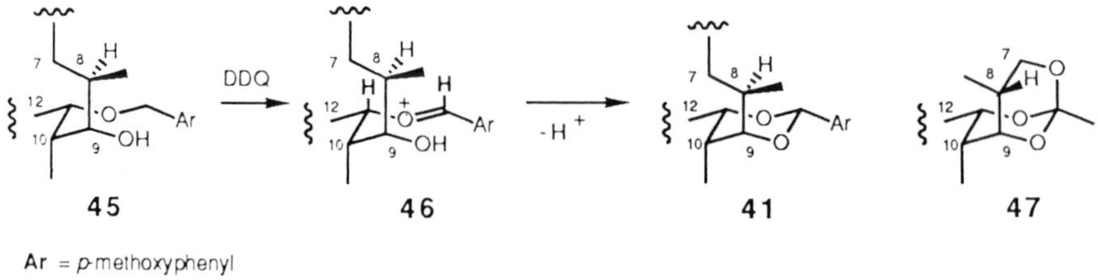

45 46 41 47

Ar = *p*-methoxyphenyl

Scheme 20

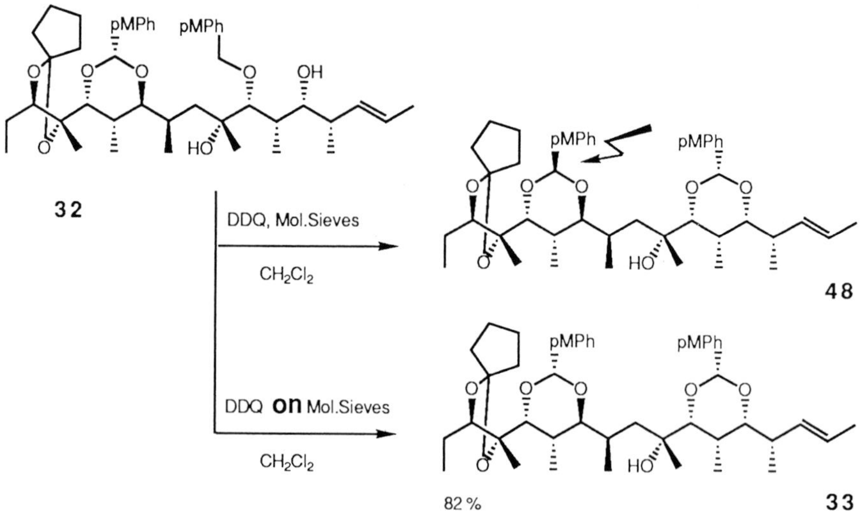

32 48 33

DDQ, Mol.Sieves

CH_2Cl_2

DDQ on Mol.Sieves

CH_2Cl_2

82 %

Scheme 21

less, the acid lability of the p-methoxybenzylidene acetal turned out to be the achilles heel of our synthesis: To initiate macrolactonization, the cyclopentylidene acetal on C12/C13 had to be cleaved while leaving the p-methoxybenzylidene acetals in place. To make sure that this is possible, model experiments were carried out early on.

We were frustrated all the more by the finding that the seco-acid **49** turned out to be quite labile in solution. As soon as the solution was concentrated de-

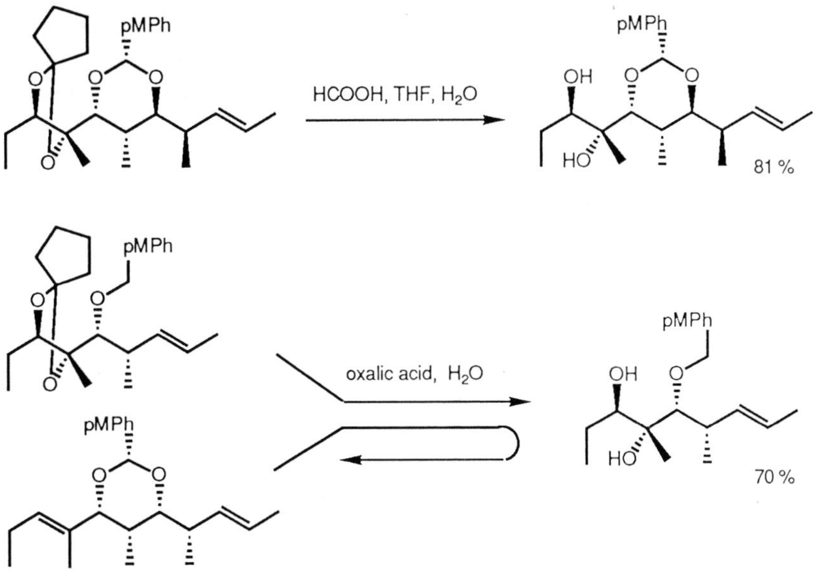

HCOOH, THF, H_2O

81 %

oxalic acid, H_2O

70 %

Scheme 22

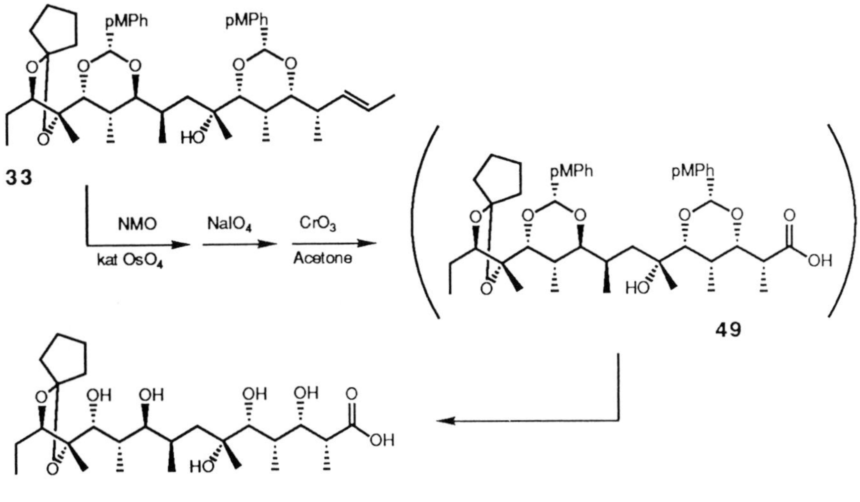

33

NMO, kat OsO_4; $NaIO_4$; CrO_3, Acetone

49

Scheme 23

composition ensued in which the p-methoxybenzylidene acetals were cleaved leaving the cyclopentylidene acetal intact.

In any case, protective groups are a disgrace to an efficient synthesis. It is even more frustrating to realize that we chose the wrong protective group in the second step of the synthesis. To start over with a different protective group on C12/C13 was therefore out of the question. Rather our attention focussed on possibilities for stabilizing the p-methoxybenzylidene acetals selectively against hydrolysis. The acid lability of the p-methoxybenzylidene acetal is caused by the stabilization of the intermediary benzyl cation by the p-methoxy group, i.e. by the electron rich aromatic system. This led us to the idea to deplete temporarily the electron density of the aromatic system by formation of a charge transfer complex. For this reason we added 10 equivalents of trinitrotoluene to the seco acid **49**. This additive fortunately opened the way for a selective hydrolysis of the cyclopentylidene acetal and allowed the successful completion of our erythronolide-A synthesis.

It was alluded to in the introduction, that there is no actual need for a synthesis of erythronolide. The reason for carrying out this synthesis was to convince

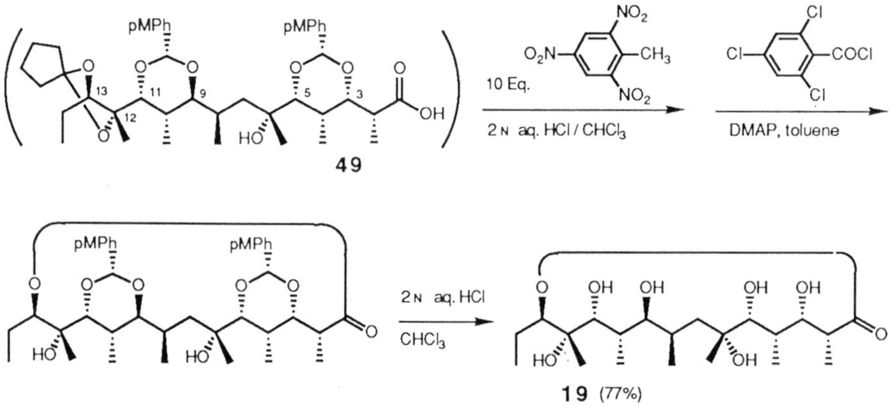
pMPh
pMPh
49
10 Eq.
2 N aq. HCl / CHCl3
DMAP, toluene
2 N aq. HCl
CHCl3
19 (77%)

Scheme 24

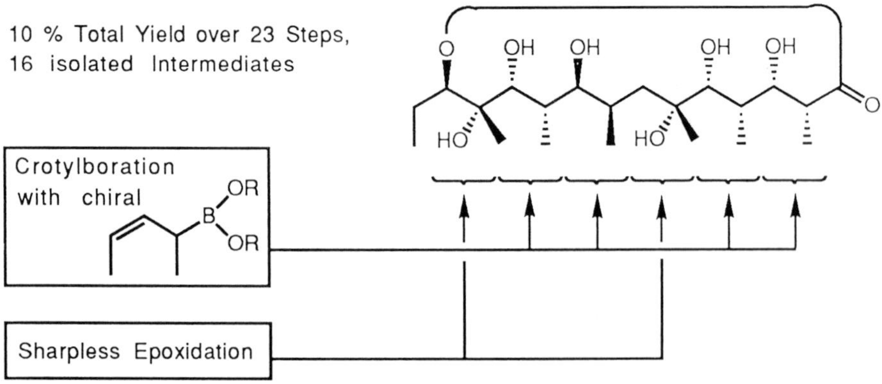
10 % Total Yield over 23 Steps,
16 isolated Intermediates
Crotylboration
with chiral
Sharpless Epoxidation

Scheme 25

ourselves of the reliability of the stereoselective chain extension reactions based on chiral pentenylboronates.

This particular reaction allowed us to carry out the hitherto shortest synthesis of erythronolide A. The efficiency of this synthesis is clearly a consequence of both the stereoselective chain extension processes as well as the advantageous use of p-methoxybenzyl protective groups.

References

[1] G. Büchi, W. D. MacLeod jr., *J. Am. Chem. Soc.* **1962**, *84*, 3205.

[2] M. Dobler, J. D. Dunitz, B. Gubler, H. P. Weber, G. Büchi, J. Padillo O, *Proc. Chem. Soc. (London)* **1963**, 383.

[3] A. Eschenmoser, *Naturwiss.* **1974**, *61*, 513.

[4] H. Muxfeldt, W. Rogalski, *J. Am. Chem. Soc.* **1965**, *87*, 933.

[5] S. H. Bertz, *J. Am. Chem. Soc.* **1982**, *104*, 5801.

[6] J. B. Hendrickson, *J. Am. Chem. Soc.* **1975**, *97*, 5784.

[7] E. Winterfeldt, *Prinzipien und Methoden der Stereoselektiven Synthese*, F. Vieweg & Sohn, Verlagsgesellschaft, Braunschweig, **1988**, S.1.

[8] 8a) G. H. Posner, *Chem. Rev.* **1986**, *86*, 831.
8b) H. Waldmann, *Nachr. Chem. Tech. Lab.* **1992**, *40*, 1133.
8c) L. E. Overman, M. M. Abelman, D. J. Kucera, V. D. Tran, D. J. Ricca, *Pure. Appl. Chem.* **1992**, *64*, 1813. - 8d) L. F. Tietze, U. Beifuss, *Angew. Chem.* **1993**, *105*, 137.

[9] W. Bartmann and B. M. Trost, *Selectivity - a Goal for Synthetic Efficiency*, Verlag Chemie, Weinheim, **1984**.

[10] S. F. Martin, D. E. Guinn, *Synthesis* **1991**, 245.

[11] R. B. Woodward, in *Perspectives in Organic Synthesis* (Edit.: A. Todd) Interscience, London, **1956**, S. 160.

[12] E. J. Corey, S. Kim, S. Yoo, K. C. Nicolaou, L. S. Melvin, jr., D. J. Brunelle, J. R. Falck, E. J. Trybulski, R. Lett, P. W. Sheldrake, *J. Am. Chem. Soc.* **1978**, *100*, 4620.

[13] R. B. Woodward et al., *J. Am. Chem. Soc.* **1981**, *103*, 3210.

[14] J. Mulzer, *Angew. Chem.* **1991**, *103*, 1484; *Angew. Chem. Int. Ed. Engl.* **1991**, *30*, 1452.

[15] R. Stürmer, K. Ritter, R. W. Hoffmann, *Angew. Chem.* **1993**, *105*, 112.

[16] P. F. Wiley, K. Gerzon, E. H. Flynn, M. V. Sigal jr., O. Weaver, U. C. Quarck, R. R. Chauvette, R. Monahan, *J. Am. Chem. Soc.* **1957**, *79*, 6062.

[17] J. Staunton, *Angew. Chem.* **1991**, *103*, 1331; *Angew. Chem. Int. Ed. Engl.* **1991**, *30*, 1302.

[18] C. H. Heathcock, J. P. Hagen, S. D. Young, R. Pilli, D.-L. Bai, H.-P. Märki, K. Kees, U. Badertscher, *Chem. Scripta* **1985**, *25*, 39.

[19] S. Masamune, M. Hirama, S. Mori, S. A. Ali, D. S. Garvey, *J. Am. Chem. Soc.* **1981**, *103*, 1568.

[20] a) C. H. Heathcock, S. D. Young, J. P. Hagen, R. Pilli, U. Badertscher, *J. Org. Chem.* **1985**, *50*, 2095.
b) S. Hoagland, Y. Morita, D. L. Bai, H.-P. Märki, K. Kees, L. Brown, C. H. Heathcock, *J. Org. Chem.* **1988**, *53*, 4730.

[21] I. Paterson, D. D. P. Laffan, D. J. Rawson, *Tetrahedron Lett.* **1988**, *29*, 1461.

[22] R. W. Hoffmann, *Angew. Chem.* **1987**, *99*, 503; *Angew. Chem., Int. Ed. Engl.* **1987**, *26*, 489.

[23] R. W. Hoffmann, H.-J. Zeiß, *J. Org. Chem.* **1981**, *46*, 1309.

[24] S. Masamune, W. Choy, J. S. Petersen, L. R. Sita, *Angew. Chem.* **1985**, *97*, 1; *Angew. Chem. Int. Ed. Engl.* **1985**, *24*, 1.

[25] R. W. Hoffmann, K. Ditrich, G. Köster, R. Stürmer, *Chem. Ber.* **1989**, *122*, 1783.

[26] J. L. Broeker, R. W. Hoffmann, K. N. Houk, *J. Am. Chem. Soc.* **1991**, *113*, 5006.

[27] 27a) M. Nakata, M. Arai, K. Tomooka, N. Ohsawa, M. Kinoshita, *Bull. Chem. Soc. Jpn* **1989**, *62*, 2618. - 27b) N. K. Kochetkov, A. F. Sviridov, M. S. Ermolenko, D. V. Yashunsky, V. S. Borodkin, *Tetrahedron* **1989**, *45*, 5109.
[28] D. Seebach, R. Imwinkelried, T. Weber, *EPC Syntheses with C,C Bond Formation via Acetals and Enamines*, in *Modern Synthetic Methods* (Edit.: R. Scheffold) Springer, Berlin, Vol. 4, **1986**, S. 125-259.
[29] 29a) R. W. Hoffmann, W. Ladner, *Chem. Ber.* **1983**, *116*, 1631
29b) K. Ditrich, *Liebigs Ann. Chem.* **1990**, 789.
[30] C. H. Heathcock, S. D. Young, J. P. Hagen, M. C. Pirrung, C. T. White, D. VanDerveer, *J. Org. Chem.* **1980**, *45*, 3846.
[31] R. W. Hoffmann, *Angew. Chem.* **1992**, *104*, 1147; *Angew. Chem. Int. Ed. Engl.* **1992**, *31*, 1124.
[32] G. Stork, S. D. Rychnovsky, *J. Am. Chem. Soc.* **1987**, *109*, 1565, 6904.
[33] H. Tone, T. Nishi, Y Oikawa, M. Hikota, O. Yonemitsu, *Tetrahedron Lett.* **1987**, *28*, 4569.
[34] O. Yonemitsu, in *Organic Synthesis in Japan; Past, Present, and Future* (Edit.: R. Noyori) Tokyo Kagaku Dozin, Tokyo, **1992**, S. 557- 565.
[35] S. D. Rychnovsky, D. J. Skalitzky, *Tetrahedron Lett.* **1990**, *31*, 945.
[36] M. Hikota, H. Tone, K. Horita, O. Yonemitsu, *J. Org. Chem.* **1990**, *55*, 7.
[37] Y. Oikawa, T. Nishi, O. Yonemitsu, *Tetrahedron Lett.* **1983**, *24*, 4037.
[38] S. L. Schreiber, Z. Wang, G. Schulte, *Tetrahedron Lett.* **1988**, *29*, 4085.

Progress in the Diels/Alder Reaction Means Progress in Steroid Synthesis

G. QUINKERT AND M. DEL GROSSO

Institut für Organische Chemie der Universität Frankfurt, Niederurseler Hang,
D-60439 Frankfurt (Main)

Abstract: The case history of achiral cantharidin reveals the development of unusual conditions under which the *Diels/Alder* reaction can play the key synthetic part.

The case history of chiral (-)-norgestrel demonstrates that the *Diels/Alder* reaction is going to become one of the best controllable structural transformations for enantioselective generation of molecular chirality through catalysis. There are, however, standard syntheses without a *Diels/Alder* reaction known.

1. Evaluation of the Diels/Alder Reaction: Cantharidin

Otto Diels and *Kurt Alder* won the 1950 Nobel Prize in Chemistry for their discovery and development of the diene synthesis. This reaction would later on be called the *Diels/Alder* reaction or [4+2]-cycloaddition. 23 years before, the two chemists had demonstrated a new reaction type: the addition of appropriate systems, soon to be named dienophiles, to reaction partners containing conjugated double bonds and hence classified as conjugated dienes [1].

"*... the results of our investigation are going to play a role not only in the discussion of many a theoretically interesting question but probably also in progress from a practical point of view. The synthetic preparation of complicated materials structurally very similar to or even identical with natural products would seem to be applicable in the near future. We explicitly reserve for ourselves the application of the reaction discovered by us.*"

In the same year *F. von Bruchhausen* and *H.W. Bersch* [2] gave an account of their efforts to synthesize the therapeutically useful drug cantharidin (Fig. 1), using furan as a diene and dimethylmaleic anhydride as a dienophile. Dehydrocantharidin, aimed at directly, was expected to smoothly afford the target compound on catalytic hydrogenation. The *Diels/Alder* reaction, however, did not take place. On the contrary, when cantharidin was dehydrogenated, furan and dimethylmaleic anhydride were isolated in place of dehydrocantharidin. This was the way the two pharmacists hit upon the first example of a retro-*Diels/Alder* reaction. The experiment had been suggested by the observation that, on determining the melting temperature of the adduct obtained from furan and maleic anhydride, the compound decomposed refurnishing the educt components. The authors were taken by surprise at the relatively mild conditions which are sufficient to cause the retro-cycloaddition. They gave their publication the title *A Peculiar*

Stereoselective Synthesis
Editors: Ottow, Schöllkopf, Schulz

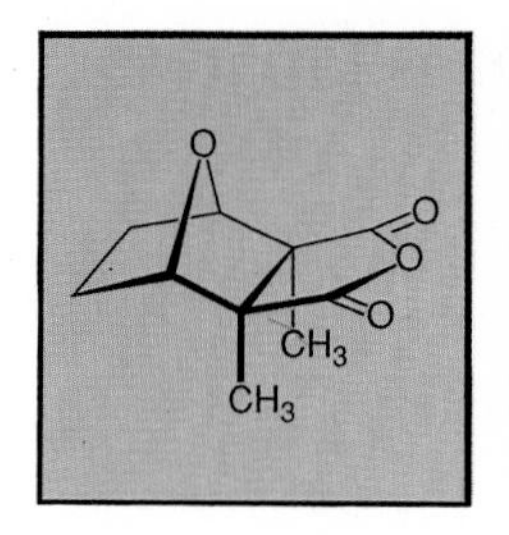

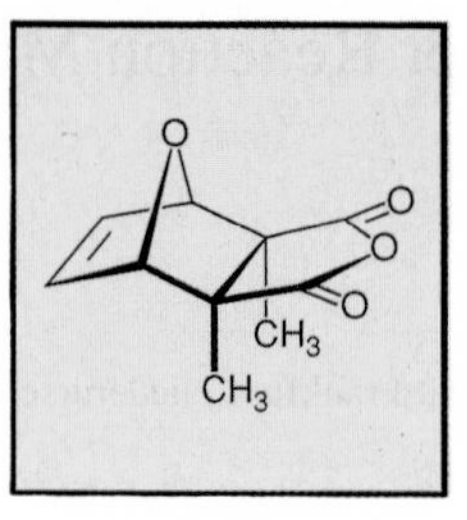

Fig. 1. Cantharidin (green field) and dehydro cantharidin (yellow field), historically important compounds testifying to the synthetical utility of the *Diels/Alder* reaction.

Decomposition of Cantharidin. They assumed not to have met a unique case and ended their paper by saying:

"We would like to ask colleagues to leave us the field of cantharidin chemistry for some time. We also intend to investigate the scope of the newly introduced decomposition reaction."

But here they obviously had lost sight of the fact that, in trying to use the *Diels/Alder* reaction for their synthesis of cantharidin, it was them who had penetrated into the hunting-ground of the Kiel chemists. The latter promptly responded [3]:

"We regret that the authors, without trying to come to an agreement with us, have published the results of their investigation which, under any interpretation, fall in the domain of problems we have explicitly reserved for ourselves. The wording of the final paragraph in our publication clearly expresses our wish and our intention, to utilize ourselves the diene synthesis found by us, above all for the construction and study of interesting natural products. It seems to us to be self-evident that someone who was fortunate in finding a reaction of widespread applicability, feels the desire to use it for solving difficult and current problems. The addition of maleic anhydride to furan is an example among cases listed by us in a patent application of 1927. In spite of their publication we cannot acknowledge priority to Messrs. v. Bruchhausen and Bersch and are not led to give up trying to synthesize cantharidin for which we have paved the way."

Rather than further follow the quarrel about territorial rights, we are interested in finding an answer to the question of whether the *Diels/Alder* reaction plays a role in biological synthesis. In their first publication *Diels* and *Alder* give the impression that they entertain that view:

"It is probable that the formation of terpenes and other natural products takes place in the same way as we have shown to happen for a whole series of examples in the laboratory."

Diels and *Alder* afterwards, one in 1939 and the other in 1953, would answer the question: *Does nature build up its products by using the diene synthesis*? *Diels* in 1939 [4]:

"In my opinion, one comes close to the truth in explaining the existence of a large number of natural products showing isoprene conjunction by assuming a

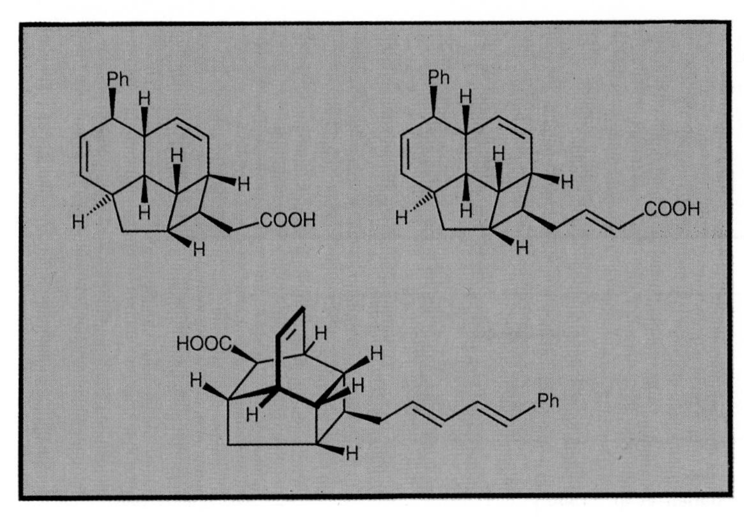

Fig. 2. Racemic mixtures of endiandric acids A-C shown to occur in nature as constituents of Endiandra species and expected to arise from achiral precursors in nonenzymatic reactions.

combination of diene synthesis with aldol condensation: two reactions that, because of the ease of their occurrence, play a dominant role in Nature."

Alder's answer [5] sounds more moderate. He believes that experience gathered meanwhile rather favours scepticism, particularly, if the case history of cantharidin is considered. Here *Diels*, in a publication from 1940 [6], had reported that all efforts to synthesize that compound so far had failed.

In pursuing the role of the *Diels/Alder* reaction in the biosynthesis of natural products, the endiandric acids [7] provide the point of the story. Fig. 2 shows the structural formulae of three constituents of Endiandra species. Despite the presence of eight stereogenic centres in each structure, all three acids are optically inactive. They happen to exist as racemic mixtures and were suspected to arise from achiral precursors in a cascade of non-enzymatic reactions. Among those reactions an intramolecular *Diels/Alder* cycloaddition was thought to take place. *K.C. Nicolaou* [8] was able to verify the endiandric acid-cascade in the laboratory and to support the existence of the non-enzymatic biosynthetic pathway.

The exotic example of the endiandric acids only underlines that the *Diels/Alder* reaction during the course of material evolution never got a chance to win. One of the most important reactions for the construction of structurally complex natural compounds proves to be the chemist's invention. But let's go back to cantharidin.

In 1980 *W.G. Dauben* et al. [9] drew attention to the fact that they had succeeded in synthesizing cantharidin, essentially in line with the approach of v. Bruchhausen and Bersch. Furan (**1**) and a latent version **2** of dimethyl maleic anhydride afforded a mixture of **3** and **4** as the major and minor components respectively (Fig. 3). After removal of the minor adduct and catalytic hydrogenolysis of the major adduct, cantharidin was isolated in high yield. To make the *Diels/Alder* reaction go reliably, high pressure conditions had to be used.

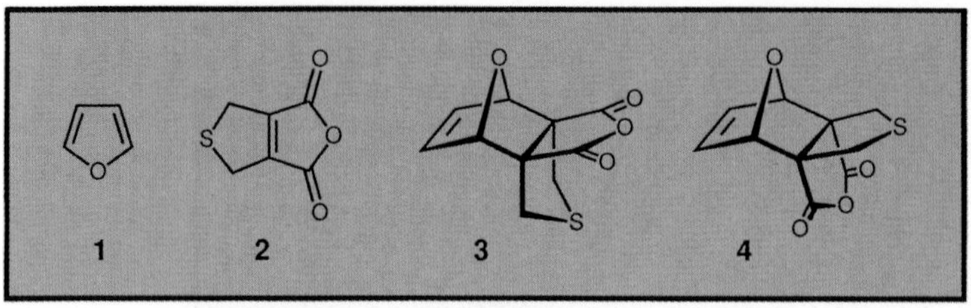

1 + 2	→	3	+	4
15 kbar, CH_2Cl_2, 6h	quant.	85	:	15
$LiClO_4$ (5M), Et_2O, 9.5h	70%	85	:	15

W.G. Dauben et al. 1980
P.A. Grieco et al. 1990

Fig. 3. Reaction conditions (yellow field) under which furan and a dienophile (green field) with the masked structure of dimethylmaleic anhydride afford stereoisomeric adduct components.

Recently *P.A. Grieco* et al. [10] have reported that the two adduct components **3** and **4** are formed in practically the same ratio, even at room temperature and under normal pressure, when **1** and **2** are dissolved in 5 M lithium perchlorate in ether (Fig. 3).

The case history of cantharidin conforms to a long standing expectation, namely that flexibility in preparative chemistry improves the chances of target-directed synthesis. Before *Dauben* or *Grieco* could successfully apply their exceptional reaction conditions for a short synthesis of cantharidin, *G. Stork* et al. [11] and *G.O. Schenck*, *K. Ziegler* [12] had accomplished this goal; each of them in a multi-step synthesis with a *Diels/Alder* reaction placed in a strategically crucial position.

As far as the synthetic utility of the *Diels/Alder* reaction is concerned, there is another candidate available: (-)-norgestrel.

2. Evaluation of the *Diels/Alder* Reaction: Norgestrel

Norgestrel (yellow field of Fig. 4) is straightforwardly connected, by the concept of latent functionality in Organic Synthesis [13], with 18a-homoestrone methyl ether (at the left hand side of the red field): the anisole part of the latter compound is a latent cyclohexenone, while the keto group conceals the ethynylcarbinol fragment. The dehydro compounds (red field) satisfy the requirement as well as 18a-homoestrone methyl ether: they all on reduction by solvated electrons go over into the hydroxy compound (green field) which after Oppenauer-oxidation is easily transformed into norgestrel. The overall yield of the structural changes shown in Fig. 4 is higher than 60%.

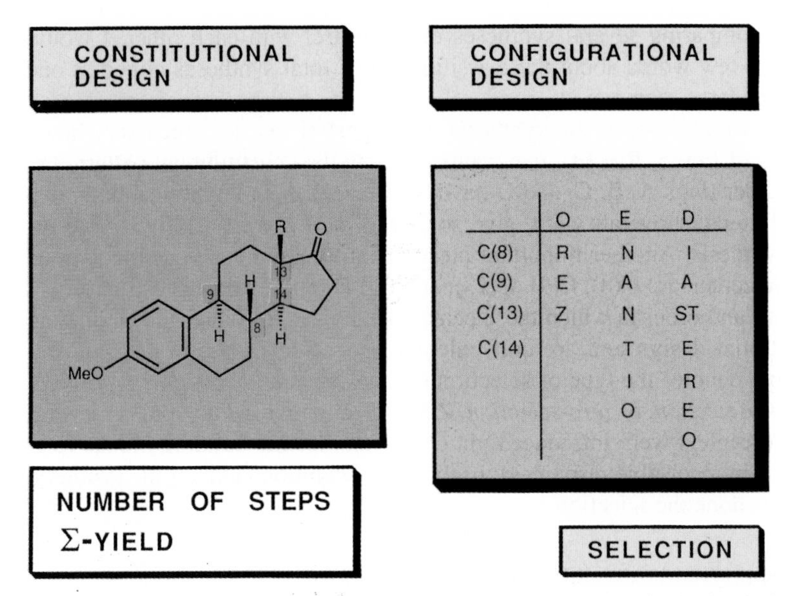

Fig. 4. The endgame of the total synthesis of (-)-norgestrel (yellow field) via 18a-homoestrone methyl ether and its 9,11- or 8,9-dehydro derivatives (red field).

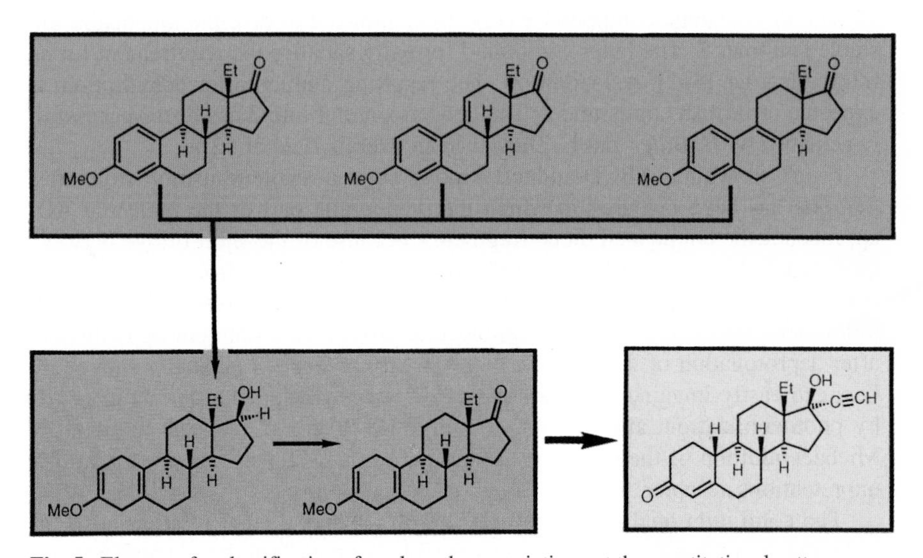

Fig. 5. Elements for classification of total syntheses pointing out the constitutional pattern, emphasizing numerical overall data (white field), and using a configurational matrix (orange field) to indicate in which order and by what type of stereoselection the stereogenic centers of the target structure (green field) have been generated.

Before comparing several syntheses of norgestrel with each other, I would like to say a few words about the classification of total syntheses aimed at one and the same target compound.

In our special case, in the synthesis of norgestrel via the precursor shown (green field of Fig. 5: R = Et), one wants to know the constitutional pattern, i.e. in which order rings A, B, C, and D have been established. Numerical data, like number of steps or overall yield, give some hint at the practicability of the respective synthesis. Another important piece of information concerns the various stereogenic centers [at C(8), C(9), C(13), and C(14) for the pretarget structure].

The way and order in which these centers have been introduced, describe the configurational design and are conveniently expressed by the orange matrix. Such a table denotes the type of selection that has been at work: *diastereoselection*, *enantioselection*, or *zero-selection*. Zero-selection means that one or several stereogenic centers were introduced via chiral-nonracemic building blocks (e.g. available from renewable resources). In this case evolution, and not the synthetic chemist, has done the selection.

2.1 Norgestrel Synthesis with Intramolecular Diels/Alder Reaction

In the first synthesis to be discussed [14] an intramolecular *Diels/Alder* reaction is the key reaction to construct the steroid skeleton, according to the constitutional pattern A+D → AD →ABCD.

The intermediate compound **1** (Fig. 6) is enolized to give the kinetically unstable tautomer **2**. The latter compound formally satisfies the requirement for an intramolecular [4+2]-cycloaddition. The resulting adduct, after dehydration, is expected to furnish compound **3**. The conversion of **1** into **3** has been successfully achieved by *H. Baier* and *G. Dürner* in an overall yield of 52%.

Fig. 7 shows the ABCD-adduct. Among eight possible transition structures, that one has been favoured in which the dienophilic part of the molecule AD' approaches the conjugated diene fragment - because of the short bridge between C(9) and C(13) - in *exo*-fashion and from that side remote from substituent R. Although the dienophilic group is electronically not at all activated, the cycloaddition does take place. As an intramolecular process it is both entropically and, after deprotonation of the dienolic hydroxy group, also enthalpically supported. One can easily imagine that compound AD' is available from its tautomer AD by photoenolization, and that AD, being a 1,5-diketone, may be obtained by Michael addition of the enantiomerically pure donor D to the stereostructurally unpretentious acceptor A.

The configurational matrix reveals that the absolute configuration at C(14) in D, due to an auxiliary's help is responsible for the diastereoselective generation of the stereogenic center C(13) in the Michael adduct AD, whilst the [4+2]-cycloaddition diastereoselectively takes care of stereogenic centers C(8) and C(9).

The auxiliary mentioned above is easily obtainable from (+)-(R)-pulegone according to *E.J. Corey* [15] and finds itself as the alcohol component in the malonic diester **2** (Fig. 8). By a tandem reaction sequence of inter- and intramol-

A + D → AD → ABCD

MeO Me Et O H 1

MeO O H Et O H 2

MeO Et O H H 3

H. Baier
G. Dürner
1985

1 —[2]→ 3
52%

Fig. 6. Crucial part of a total synthesis of (-)-norgestrel following the constitutional pattern A+D → AD → ABCD including photoenolization of **1** and intramolecular *Diels/Alder* reaction of **2** followed by dehydration furnishing steroid **3** in an overall yield of 52 %.

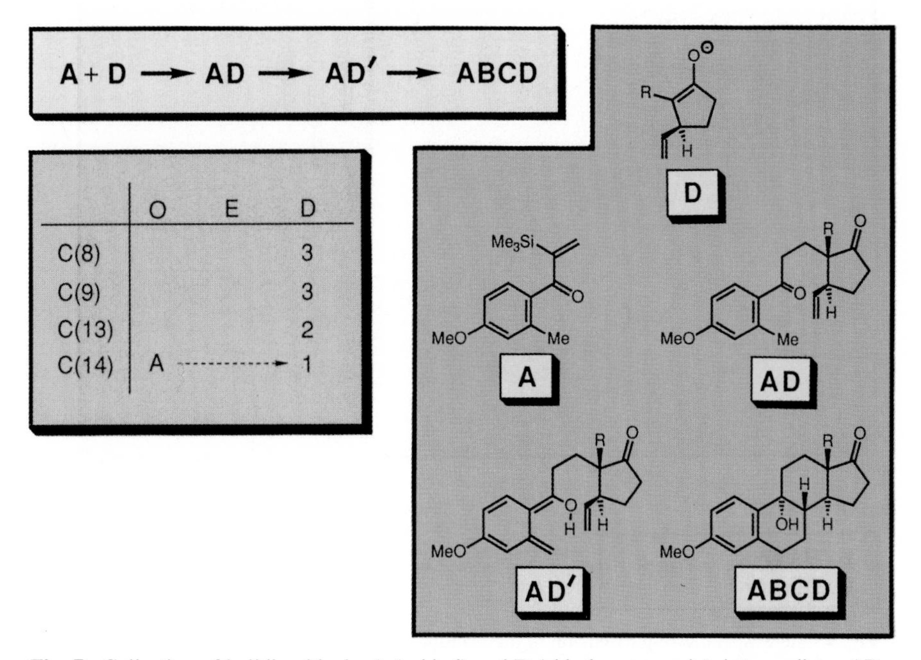

	O	E	D
C(8)			3
C(9)			3
C(13)			2
C(14)	A -----------→		1

Fig. 7. Collection of building blocks A (achiral) and D (chiral nonracemic), intermediates AD and AD', and main component of *Diels/Alder* reaction with steroid skeleton ABCD.

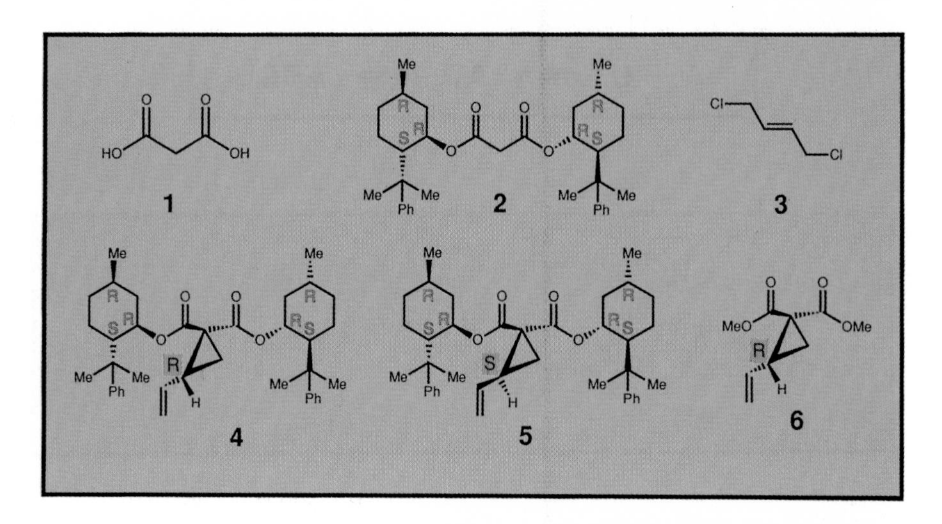

U. Schwartz 1982

1 —83%→ 2, 3 —67%→ 4, 5; 4 —65%→ 6

Fig. 8. Diastereoselective formation of the three-membered ring compound **6** utilizing the C_2-symmetrical diester **2**.

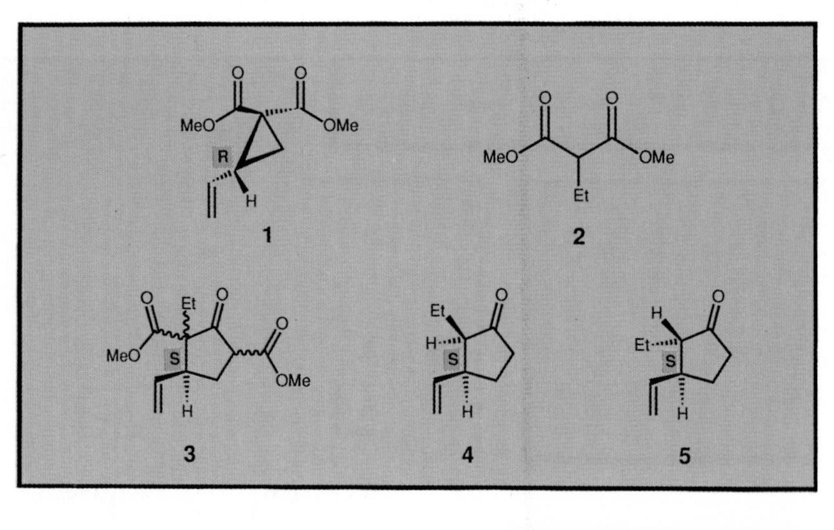

H. Baier
G. Dürner 1985

1, 2 → 3 —44%→ 4 ⇌ 5

Fig. 9. Stereospecific ring extension of the cyclopropane derivative **1** on reaction with the enolate anion of ethyl methylmalonate furnishing a mixture of the enantiomerically pure five-membered ring compounds **4** and **5**.

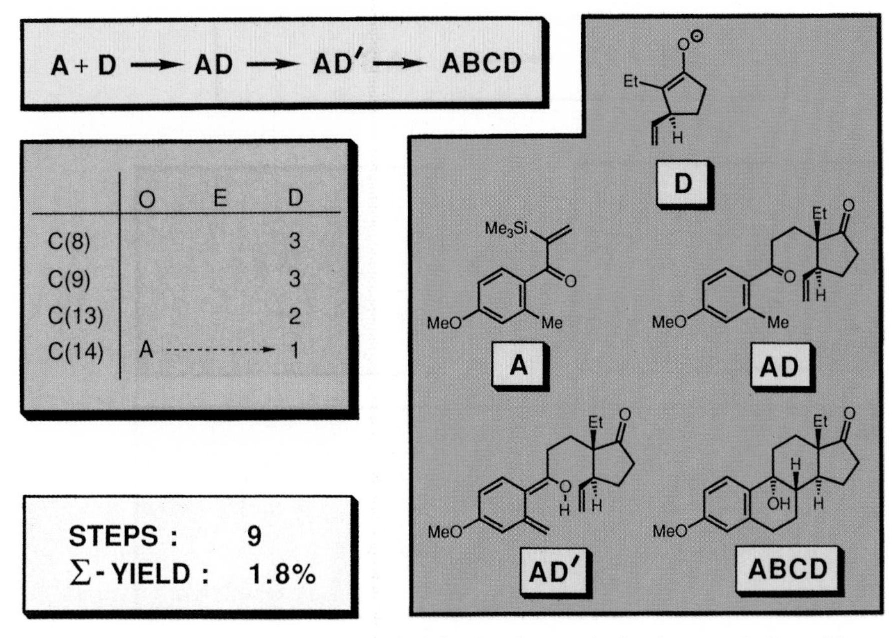

Fig. 10. Evaluation of the steroid synthesis following the constitutional pattern A+D → AD → AD′ → ABCD which gives (-)-norgestrel after nine steps in an overall yield of 1.8% (relative to malonic acid).

ecular substitutions, starting from **2** and **3**, a mixture of diastereomers **4** and **5**, in a ratio of 98:2, is produced [16]. After **5** has been separated, transesterification converts **4** into **6**. The overall yield of **6** with R-configuration at the stereogenic center amounts to 36% relative to malonic acid **1**.

S. Danishefsky [17] has shown, how to effect ring expansion of the three-membered ring compound **1** (Fig. 9) starting with inversion displacement at the stereogenic center. When **1** was treated with the enolate anion of dimethyl ethyl malonate (**2**), a product was obtained which, after hydrolysis and decarboxylation, furnished a mixture of diastereomers **4** and **5** in 44% overall yield. Under the appropriate conditions, both the five-membered ring compounds give the Michael donor D of Fig 7.

To sum up: (-)-norgestrel has been synthesized as outlined (Fig. 10) after 9 steps in an overall yield of 1.8%. The individual steps are extremely stereoselective. The way to the enantiomerically pure building block D, however, is rather tedious.

2.2 Norgestrel Synthesis with Intermolecular *Diels/Alder*-Reaction

Structure **2** (Fig. 11), retrosynthetically accessible from structure **1** by dehydrogenation, could be thought of as being a *Diels/Alder* adduct, if only rings C and

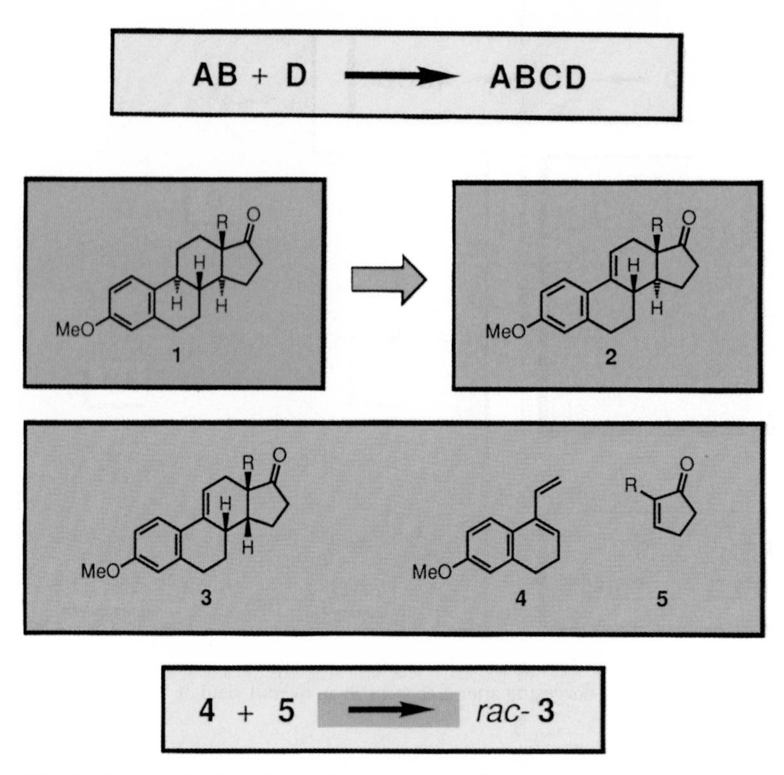

Fig. 11. Retrosynthetic analysis of structure **1**: As far as the constitution is concerned structure **2** correlates by dehydrogenation with **1** or by retro *Diels/Alder* reaction with **4** and **5**. As far as the configuration is concerned **4** and **5** correlate by *Diels/Alder* reaction with **3**, not with **2**.

D were cis-fused. In point of fact, the *Dane* diene **4** and cyclopentenone **5** (R = Me or Et) on reaction are expected to lead to rac-**3**. Since this [4+2]-cycloaddition would comfortably construct the steroid skeleton, one might accept additional effort required to correct the configuration at C(14).

As *E. Dane* et al. [18] found out in 1939, under normal conditions **4** and **5** (Fig. 12) did not react at all. Changing the dienophile from **5** to cyclopent-2-en-1-one made the reaction go. As in the case of cantharidin, the non-methylated dienophile did react. One might think of using high pressure (Fig. 3) or remember the effect of 5M lithium perchlorate in ether.

One might instead ask, whether cases are known, in which catalysts have accelerated *Diels/Alder* reactions. *Alder* in his various essays asked himself that question again and again. In his Nobel lecture he pointed out:

"... *polar solvents and catalysts have relatively little effect on diene synthesis*",

and elsewhere he said:

"... *up to now it has not been possible to decouple and rearrange the six π-electrons (sic) in the reaction complex, in order to influence the reaction catalytically...* "

In the early sixties it was reported [19] that quite a few *Diels/Alder* reactions had been successfully accelerated by Lewis acids. The limiting factor usually is poly-

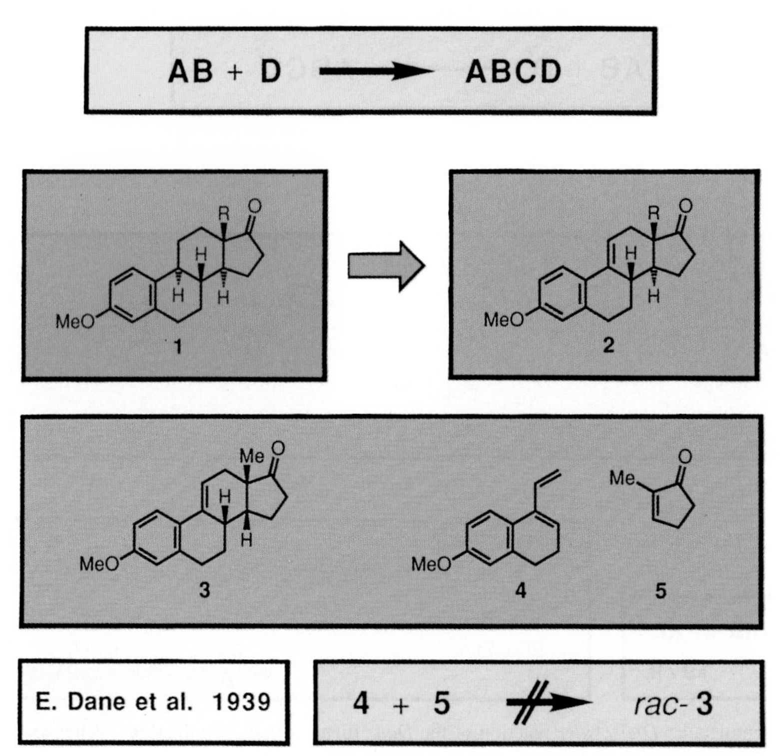

Fig 12. According to *Dane* et al. *Diels/Alder* reaction of the *Dane* diene **4** with **5** does not take place.

merization of the reactant, especially of the diene. Differential acceleration of two reactions each leading to a different adduct, can cause change in product composition. A striking example is available, which is even relevant to the matter in hand.

Z. Valenta et al. [20] in 1979 reported on the reaction of the *Dane* diene **1** (Fig. 13) and 2,6-dimethyl-1,4-benzoquinone (**2**) in boiling benzene, affording after 24 h adduct **4** in 82% yield, which has obviously got the 'wrong' constitution as far as the construction of the steroid skeleton is concerned. After 30 min at -15°C in ether, however, the adduct components **3** and **4** in a ratio of 86:14 were produced, provided $BF_3 \cdot OEt_2$ had been present. The major adduct can easily be isolated in 69% yield. Undoubtedly, the dramatic change in stereoselection promises to be of synthetic value.

The question comes up, of course, of whether adduct formation between the *Dane* diene **1** and dienophile **2** (Fig. 14) might be brought about by Lewis acid. The answer is 'yes'. Depending on the nature of the Lewis acid, the primary adduct rac-**3** or the secondary isomerization product rac-**4** can be isolated [21].

The constitutions and relative configurations of the various isolated products have been determined by X-ray analysis. Fig. 15A shows a representation of the adduct of type **3** (Fig. 14). Fig. 15B and Fig. 15C represent adducts with a double bond between C(8) and C(9) and a methyl or even an ethyl group at C(13), respectively. In all three cases, cis-fusion of rings C and D is clearly to be recognized.

AB + D′ ⟶ ABCD′

1 + 2 ⟶	*rac*-3	+ *rac*-4
uncatalyzed :	—	82%
BF_3OEt_2 :	69%	10-15%

Z. Valenta et al. 1979

Fig. 13. Intermolecular *Diels/Alder* reaction of the *Dane* diene **1** with 2,6-dimethyl-1,4-benzoquinone (**2**): direction of stereoselection is inverted in the presence of Lewis acid.

AB + D ⟶ ABCD

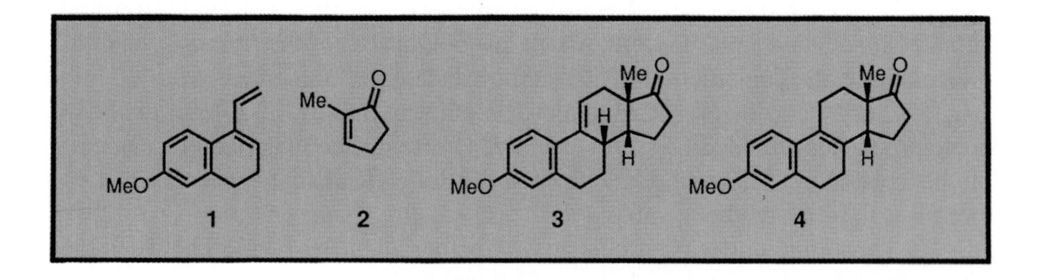

1 + 2 ⟶	*rac*-3 ⟶ *rac*-4
uncatalyzed :	no reaction
$TiCl_4$:	89% · *rac*-4
Et_2AlCl :	55% *rac*-3

E. Dane et al. 1939
M. Del Grosso 1991

Fig. 14. Lewis acids cause the *Diels/Alder* reaction of the *Dane* diene **1** with 2-methylcyclopent-2-en-1-one to occur, after all.

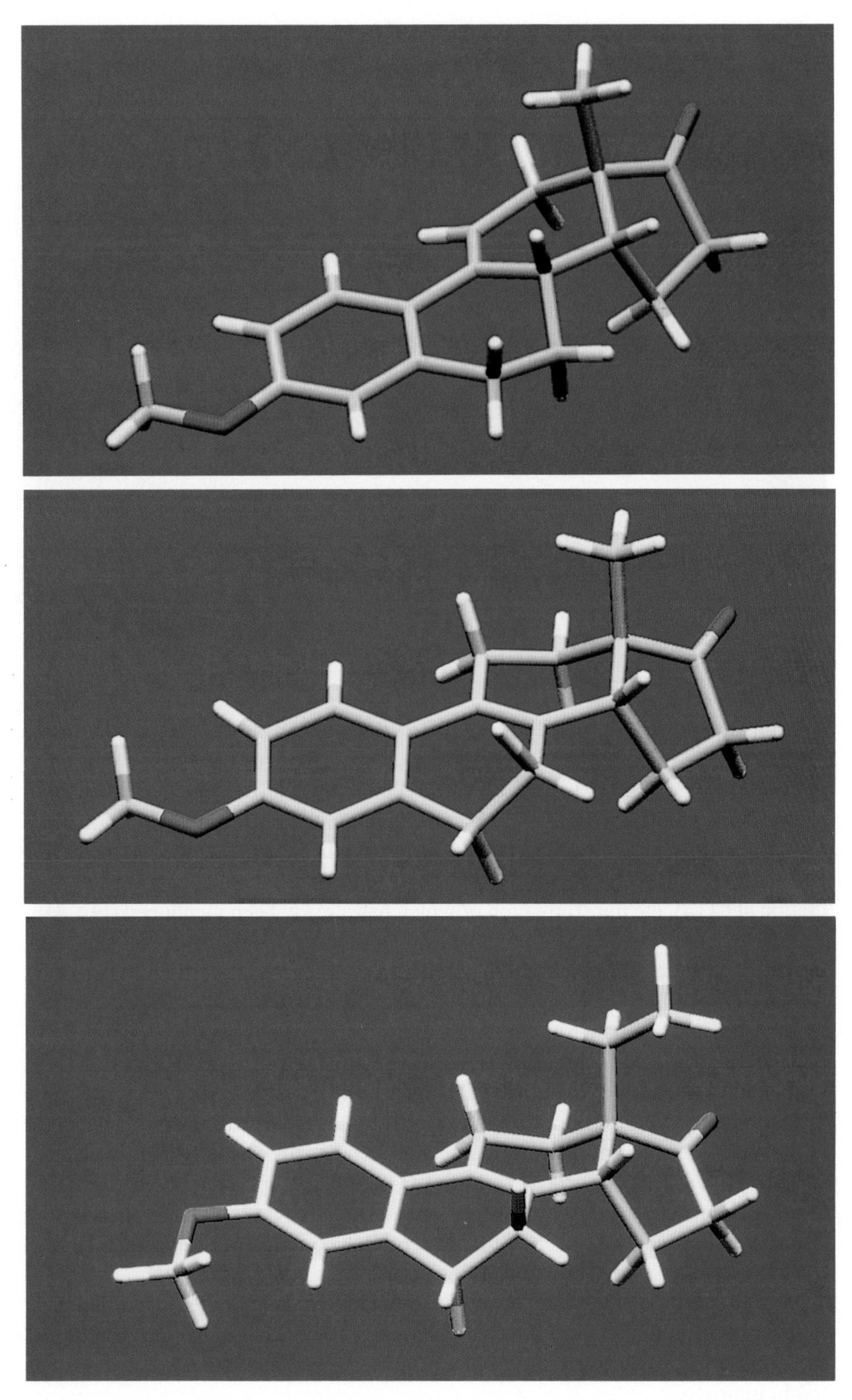

Fig. 15. Polytube representation (MacroModel [22]) of the crystal structure of diverse *Diels/Alder* adducts. A: rac- **3** (cf. Fig. 14); B: rac- **4** (cf. Fig. 14); C: the 18a-homologue of rac- **4** (cf. Fig. 14). The bonds between adjacent atoms are colour-coded: C = grey; H = white; O = red.

STEREOSTRUCTURAL CORRECTION

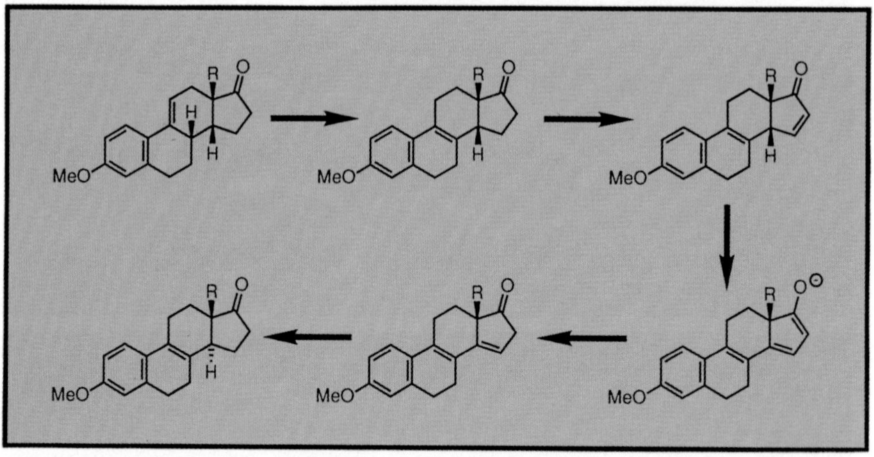

Fig. 16. How configuration at C(14) of *Diels/Alder* adducts of type **3** or **4** (cf. Fig. 14) can be corrected.

In a straightforward procedure (Fig. 16) trans-fusion of rings C and D can be established. The α,β-unsaturated ketone occupies a key position. It is produced in the usual way [23] by oxidation of the silyl enol ether of the preceding nonconjugated ketone with $Pd(OAc)_2$ [24] and converted into the β,γ-unsaturated ketone by a deprotonation/protonation sequence. Here the synthetic route merges into the *Torgov* pathway [25]: catalytic hydrogenation affords a steroid with the correct configuration at C(14), the transformation of which has already been mentioned (Fig. 4).

The black arrow on red ground in Fig. 14 denotes a *chirogenic* reaction step, by which two achiral educt components are transformed into a chiral product. *A. Eschenmoser* [26] has coined that term, to emphasize that those "specially designated reaction steps carry the opportunity for enantioselective generation of molecular chirality through catalysis".

Having effected the imaginary *Dane* reaction through Lewis acids, we were interested to find out, whether enantioselective adduct formation would take place in the presence of chiral-nonracemic ligands . The result was only marginally successful. This is not too surprising, however, as the dienophilic monoketone ought to be only loosely fixed to the central atom of the Lewis acid. This follows from X-ray structure analysis [27] of the dimeric 1:1 complex between $TiCl_4$ and 2-methylcyclopent-2-en-1-one (Fig. 17).

Would it not be advantageous to have a bidentate dienophilic ligand on hand? After complexation, a reduced conformational space for the transition structure of the *Diels/Alder* reaction would be available. In consequence, stereoselection might improve.

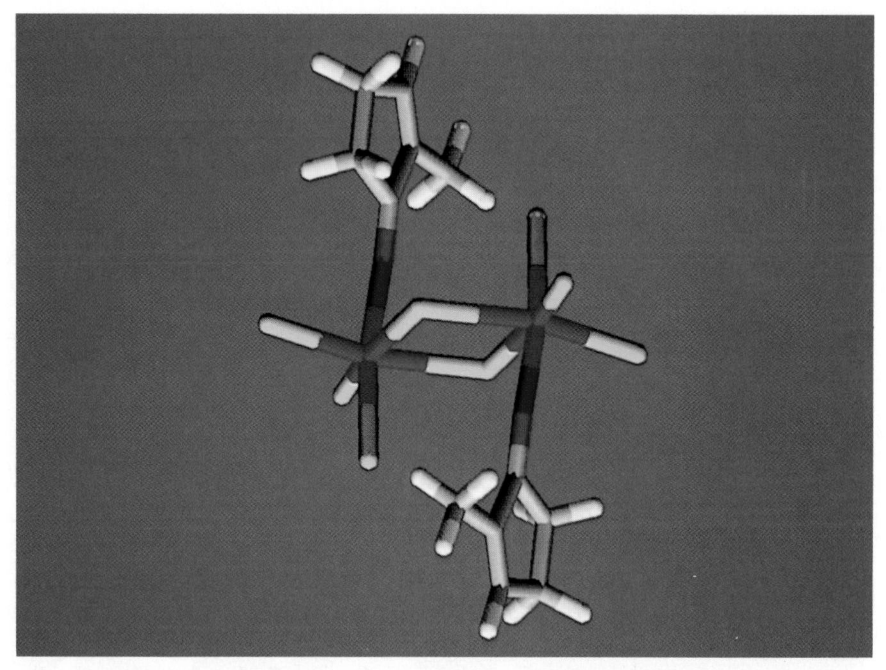

Fig. 17. Single crystal structure analysis of di-μ-chloro-2-bis[trichloro(2-methylcyclopent-2-en-1-one)titanium]. For colour code see legend of Fig. 15 and: Ti = blue, Cl = yellow.

AB + D ⟶ ABCD

MeO 1 Me O O 2 Me O H OH H MeO 3 H H OH Me O MeO 4

E. Dane et al.	1939
G. Singh	1956
A. Bucher	1990

1 + 2 ⟶ *rac*- 3 + *rac*- 4

82% 25 : 75

Fig. 18. *Diels/Alder* reaction of the *Dane* diene **1** and 3-methylcyclopent-3-en-1,2-dione (**2**) affords adducts rac- **3** and rac- **4**.

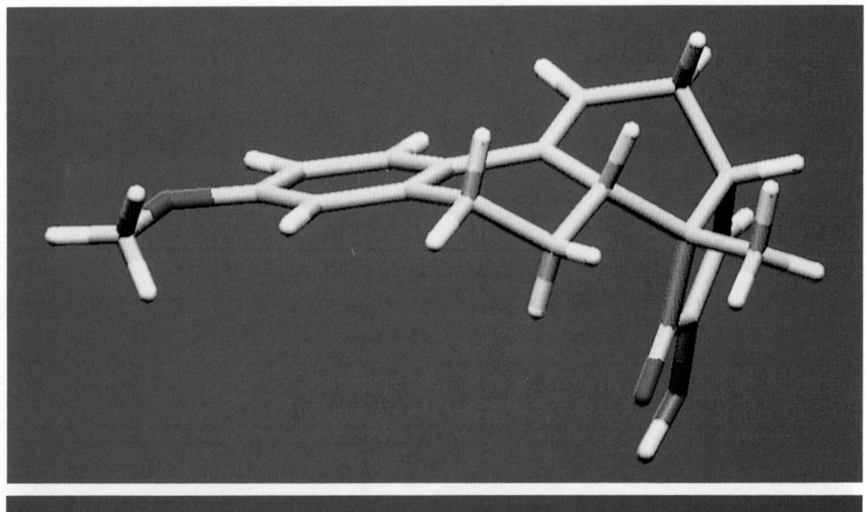

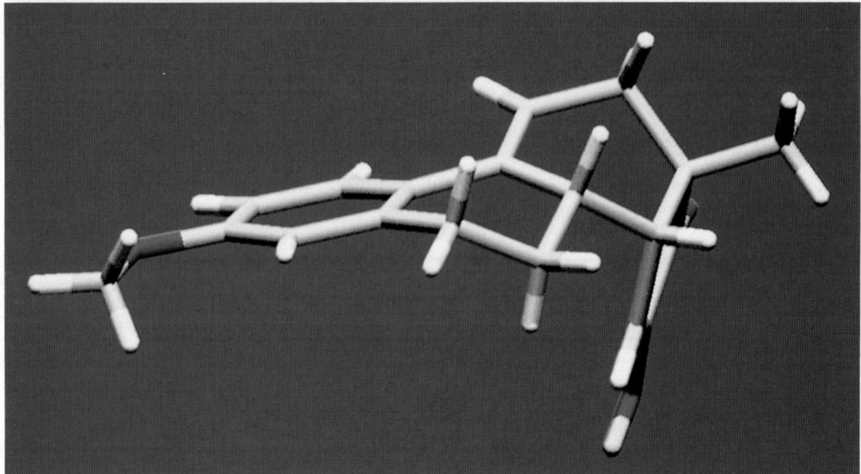

Fig. 19. Single crystal structure analysis of **3** (A) and **4** (B) (cf. Fig. 18).

3-Methylcyclopent-3-en-1,2-dione (**2**) (Fig. 18) had already been chosen as a dienophile to react with *Dane* diene **1** by the Munich group [28]. The isolated product was sent to Schering AG then for an estrogenic activity test. No positive result has ever been reported.

G. Singh [29], working in *R.B. Woodward*'s laboratory at Harvard University later on reinvestigated the *Diels/Alder* adduct. It proved to be a mixture of rac-**3** and rac-**4** as the major and minor components, respectively, and this was confirmed by *A. Bucher* and *J.W. Bats* [21] who carried out the X-ray structure determination (Fig. 19).

While the uncatalyzed reaction went astray, boron trifluoride in ether changed the direction of regioselection, furnishing the isomer with the steroidal skeleton, rac-**3** (Fig. 20), almost exclusively.

AB + D ⟶ ABCD

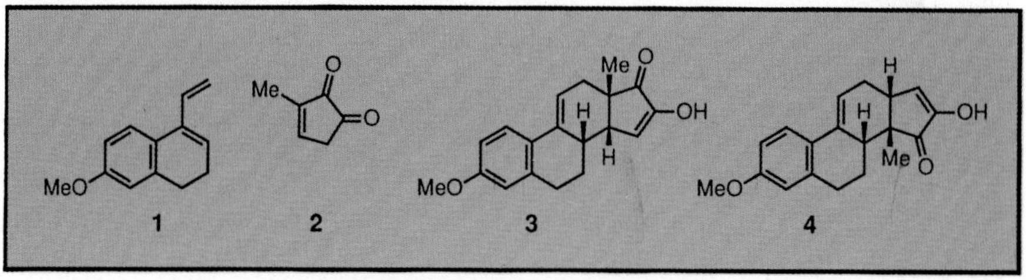

E. Dane et al.	1939
G. Singh	1956
A. Bucher	1990

1 + 2	⟶	*rac*-3	+	*rac*-4
uncatalyzed :	82%	25	:	75
BF_3OEt_2 :	75%	98	:	2

Fig. 20. *Diels/Alder* reaction of the *Dane* diene **1** and the diketonic dienophile **2**: direction of stereoselection is inverted in the presence of Lewis acids.

With the experience in hand, that the course of the *Diels/Alder* reaction between **1** and **2** can be directed, Lewis acids in which the central metal atom can easily extend its coordination number to six, were put high on the list. Unfortunately, we did not succeed in getting crystals from complexes of 3-methylcyclopent-3-en-1,2-dione **2** and Lewis acids. We were successful [30], however, in carrying out the single crystal structure determination of several complexes of other 1,2-diketones and $TiCl_4$ or $SnCl_4$ (Fig. 21).

With $TiCl_4$ or $SnCl_4$, acenaphthenquinone, benzil and camphorquinone function as bidentate ligands, while the last compound on complexation with MAD acts as a monodentate ligand. On reaction of the *Dane* diene **1** and 3-ethylcyclopent-3-en-1,2-dione (**2**) it was shown by *W. Döring* [31] (Fig. 22) that the type of Lewis acid used determines adduct composition.

$TiCl_4$ causes polymerization of the diene. $TiCl_2(OiPr)_2$ gives only rac-**5**, and $TiCl(OiPr)_3$ affords merely rac-**3**. The latter compound was identified by X-ray structure analysis (Fig. 23).

By the use of $TiCl_2(OiPr)_2$ in the presence of appropriately substituted TADDOLs [32] *M. Bauch* [33] effected the enantioselective version of the chirogenic reaction step of Fig. 24.

In one case (R^1 = phenanthren-9-yl; $R^2 = R^3$ = ethyl) compound **3** was isolated in 65% chemical yield and an enantiomeric excess of 93%. When the *Dane* diene **1** reacted with the ethyl substituted diketone **2** (Fig. 25) under the above mentioned conditions (20 mol% of the chiral-nonracemic ligand) **3** was accessible in 77% chemical yield and an enantiomeric excess of 88% [31].

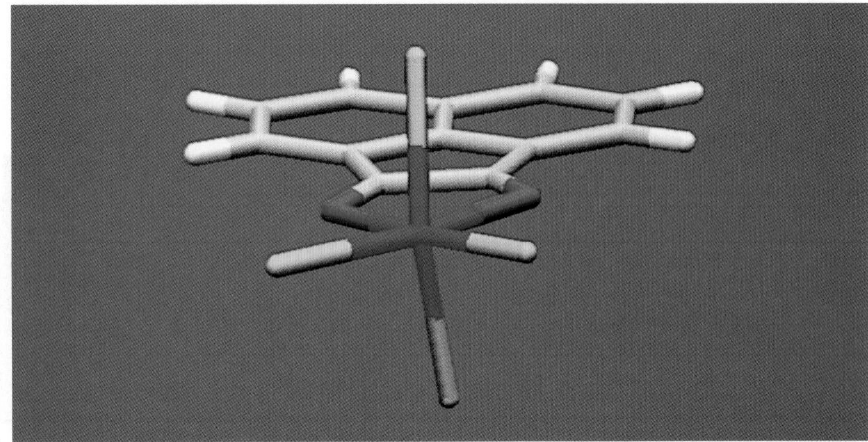
A

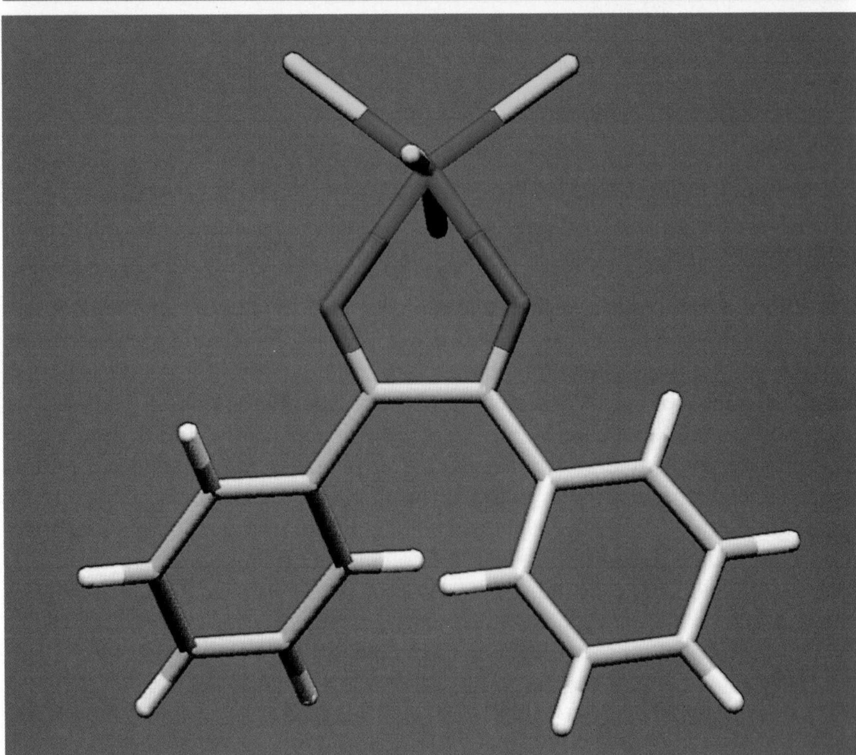
B

Fig. 21. Single crystal X-ray structure of complexes of acenaphthenquinone and $TiCl_4$ (A), of benzil and $SnCl_4$ (B), of camphorquinone and $TiCl_4$ (C), and of camphorquinone and MAD bis [2,6-bis(1,1-dimetylethyl)-4-methylphenolato]methylaluminum (D) [30].

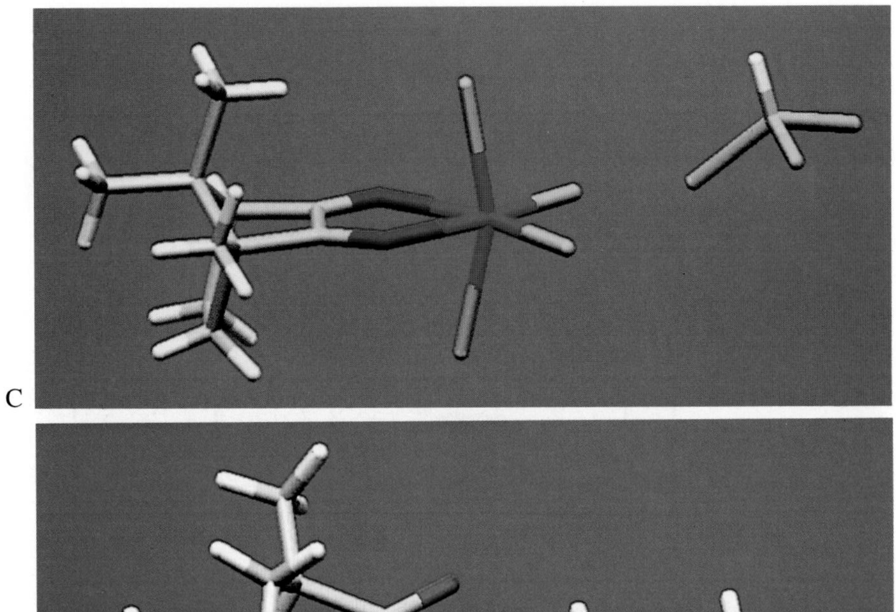
C

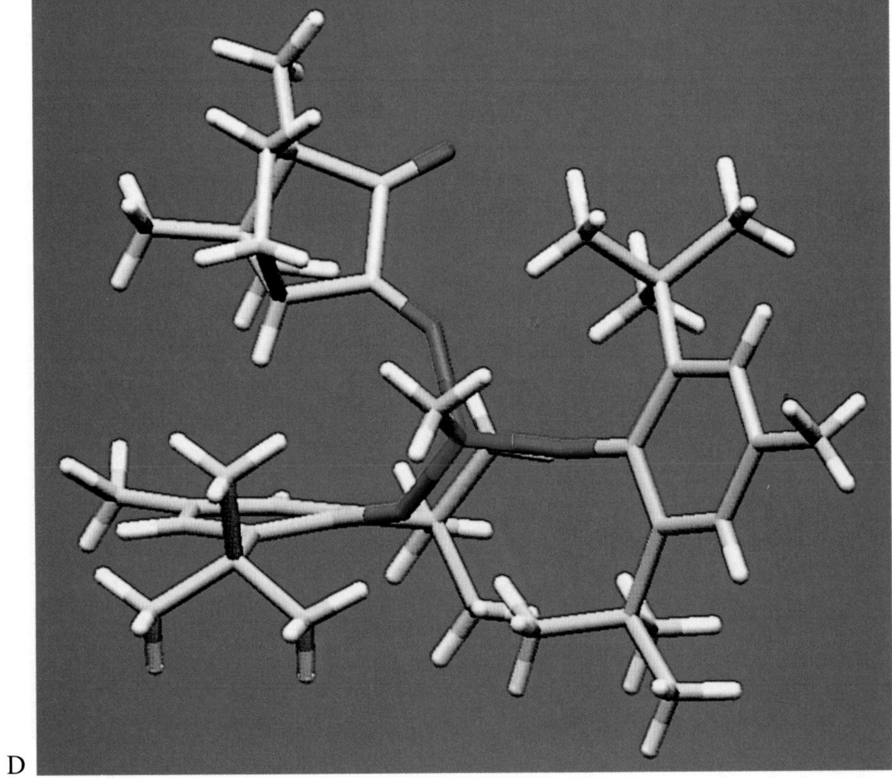
D

Fig. 21. (continued)

AB + D ⟶ ABCD

	$1 + 2$ ⟶	*rac*-3	:	*rac*-4	*rac*-5
uncatalyzed :	40%	35	:	65	—
BF_3OEt_2 :	53%	97	:	3	—
$TiCl(O\textit{i}Pr)_3$:	70%	>99.5	:	<0.5	—
$TiCl_2(O\textit{i}Pr)_2$:	80%	—		—	>99
$TiCl_4$:	—	—		—	—

W. Döring 1992

Fig. 22. *Diels/Alder* reaction of the *Dane* diene **1** and 2-ethylcyclopent-2-en-1-one (**2**): adduct composition depends on the Lewis acid used.

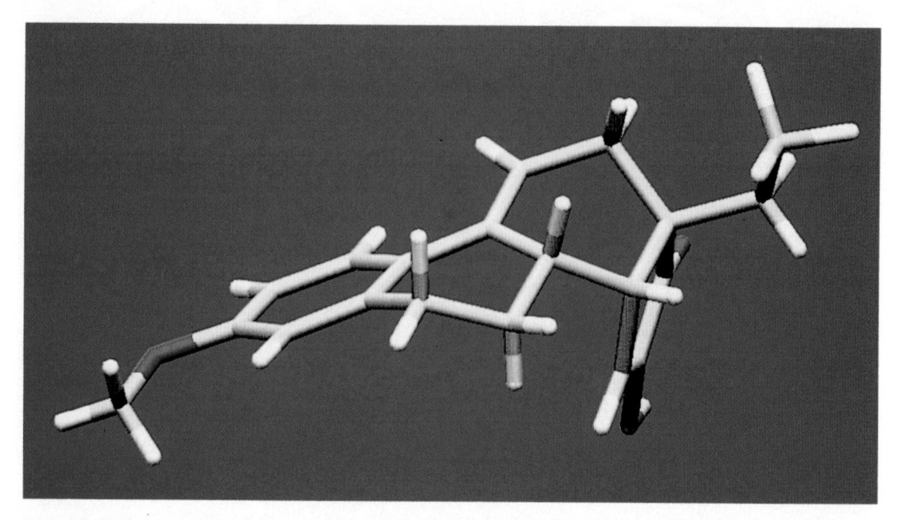

Fig. 23. Single crystal X-ray structure analysis of the *tert*-butyldimethylsilyl ether of adduct **3** (cf. Fig. 22).

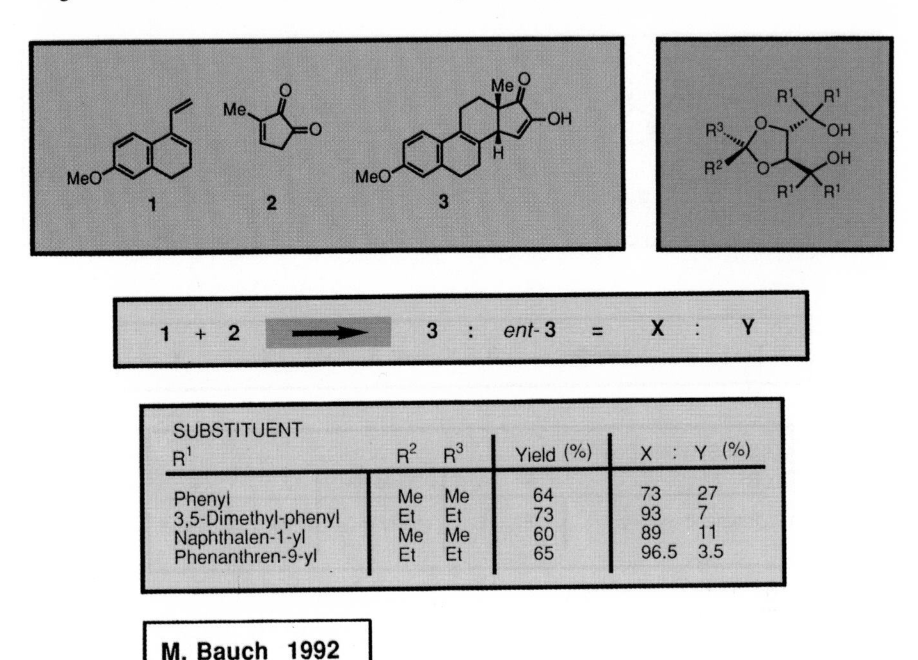

SUBSTITUENT R^1	R^2	R^3	Yield (%)	X	Y (%)
Phenyl	Me	Me	64	73	27
3,5-Dimethyl-phenyl	Et	Et	73	93	7
Naphthalen-1-yl	Me	Me	60	89	11
Phenanthren-9-yl	Et	Et	65	96.5	3.5

Fig. 24. Enantioselective verification of the chirogenic *Diels/Alder* reaction of the *Dane* diene **1** and **2** in the presence of a chiral-nonracemic Ti-complex.

The particular *Diels/Alder* adduct (R = Me or Et) after deoxygenation and isomerization was treated further using the *Torgov* route (Fig. 26). Deoxygenation was achieved by treating the triflate with tributyltin hydride in the presence of LiCl and $Pd(PPh_3)_4$ [34]. Enantioselection, which began during the [4+2]-cycloaddition, was completed by differential crystallization of the deoxygenated steroid [27].

To sum up, (+)-estrone is accessible as outlined (Fig. 27) after seven synthetic steps in an overall yield of 9.6%, using the intermolecular *Diels/Alder* reaction of the *Dane* diene AB and dienophile D in the presence of $TiCl_2(OiPr)_2$ and the auxiliary A.

Similarly, (-)-norgestrel was obtained (Fig. 28) after nine synthetic steps in an overall yield of 7.5 %. Here again, stereogenic centers C(13) and C(14) were generated enantioselectively making use of the auxiliary A as a ligand in a titanium complex. After the configuration at C(14) had been corrected, stereogenic centers C(8) and C(9), as well as C(17) and C(10), were introduced in the usual way with high diastereoselection.

It is not without appeal to compare the (-)-norgestrel synthesis of the constitutional type AB+D → ABCD with that one following the pattern A+D → AD → ABCD. The real touchstone, however, is the synthesis performed on an industrial scale. The industrial synthesis of (-)-norgestrel (Fig. 29), designed and performed at Schering AG [35], more or less follows the strategy of the *Torgov* syn-

1 + 2 → 3 : *ent*-3 = X : Y

SUBSTITUENT R^1	R^2	R^3	Yield (%)	X : Y (%)	
Phenanthren-9-yl	Et	Et	77	94	6

W. Döring 1993

Fig. 25. Enantioselective verification of the chirogenic *Diels/Alder* reaction of the *Dane* diene **1** and 3-ethyl-3-cyclopenten-1,2-dione **2** in the presence of a chiral-nonracemic Ti-complex.

STEREOSTRUCTURAL CORRECTION

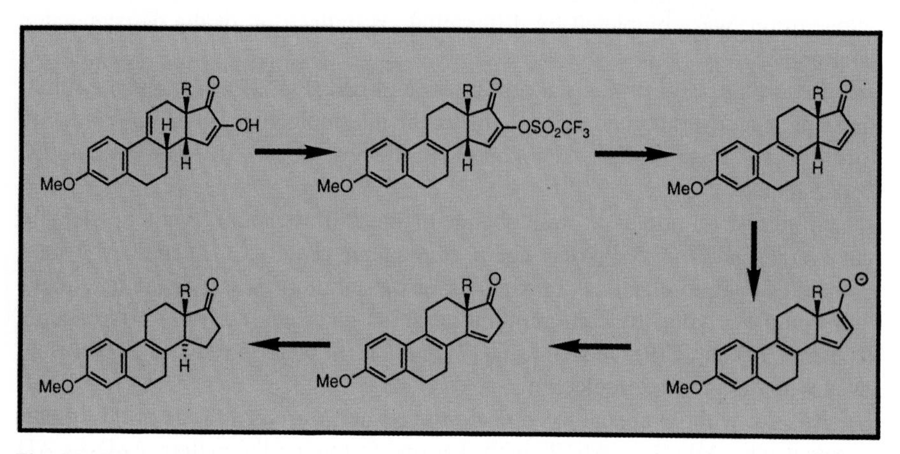

Fig. 26. How configuration at C(14) of *Diels/Alder* adducts of type **3** (cf. Figs. 24 or 25) has been corrected.

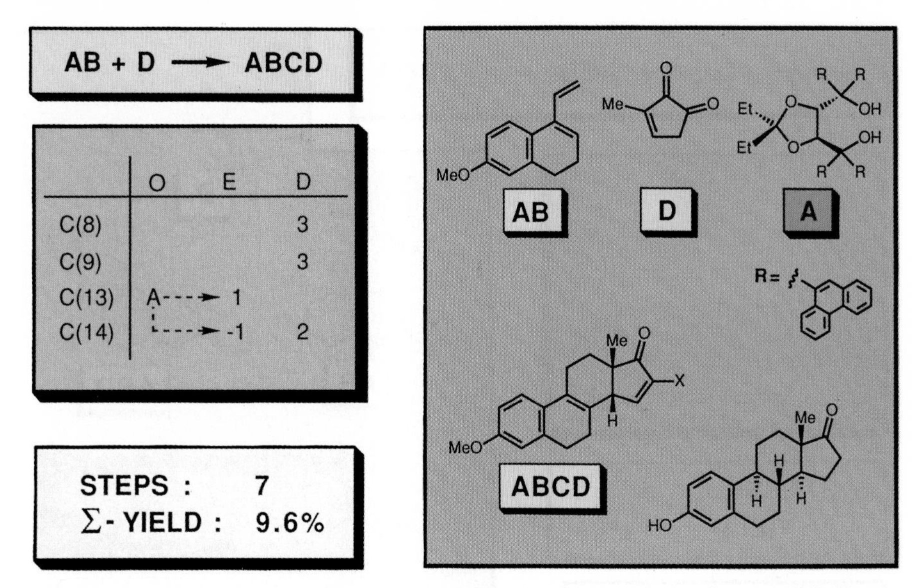

Fig. 27. Evaluation of the steroid synthesis following the constitutional pattern AB+D → ABCD which gives (+)-estrone after seven steps in an overall yield of 9.6 % (relative to building block AB).

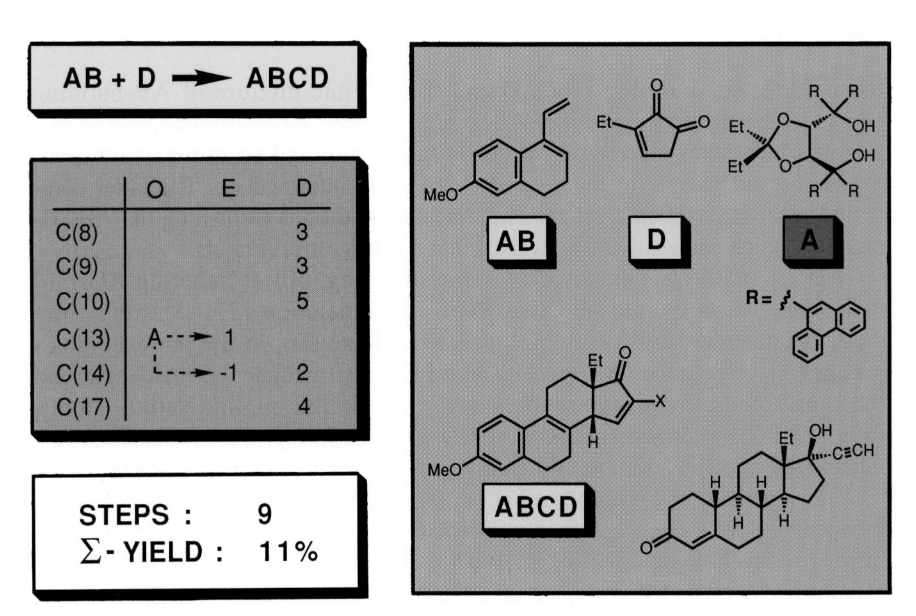

Fig. 28. Evaluation of the steroid synthesis following the constitutional pattern AB+D → ABCD which gives (-)-norgestrel after nine steps in an overall yield of 7.5 % (relative to building block AB).

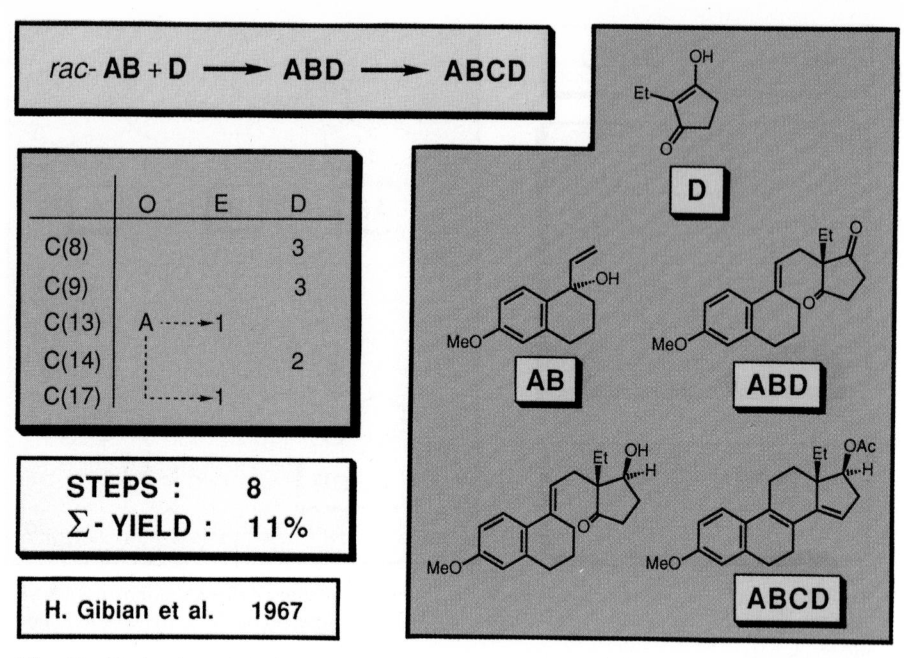

Fig. 29. Evaluation of the industrial steroid synthesis following the constitutional pattern AB+DA → BCD which gives (-)-norgestrel after eight steps in an overall yield of 11 % (relative to building block rac-AB).

thesis [25]. The building block D and the racemic mixture of AB-building blocks react affording an achiral ABD-intermediate, the microbiological reduction of which takes place with extremely high regio- and enantioselection leading almost exclusively to the desired chiral ABD-intermediate. The latter compound, after acetylation, cyclizes under acidic conditions furnishing the ABCD-product, which can easily be converted into (-)-norgestrel (Fig. 4).

The alternative synthesis of (-)-norgestrel (Fig. 30) at Schering AG [36] makes use of an enantioselective *Robinson* annelation [37]. Starting from methylvinylketone and 2-ethylcyclopenta-1,3-dione (D), in the presence of L-proline (A), one reaches the optically active CD-intermediate. Formaldehyde and the substituted ethyl acetoacetate shown provide the missing carbon atoms. From the BCD-intermediate to the ABCD-diketone is only a short distance. Partial and site specific addition of the ethynyl group completes a lucid synthesis.

This synthesis of (-)-norgestrel is one of the masterpieces of total syntheses. Its opening marks a new epoch in synthetic chemistry. There is much merit in removing a psychological barrier. In 1894 *E. Fischer* proclaimed the principle of asymmetric synthesis. Ever since then a case was to be expected, where one of two enantiomers would be formed in excess. One was not prepared, however, to find out, more than 75 years later, that this would happen under relatively simple conditions and in the course of a development of an industrial synthesis. In the history of chemistry this breakthrough will be connected forever with the name of *R. Wiechert.*

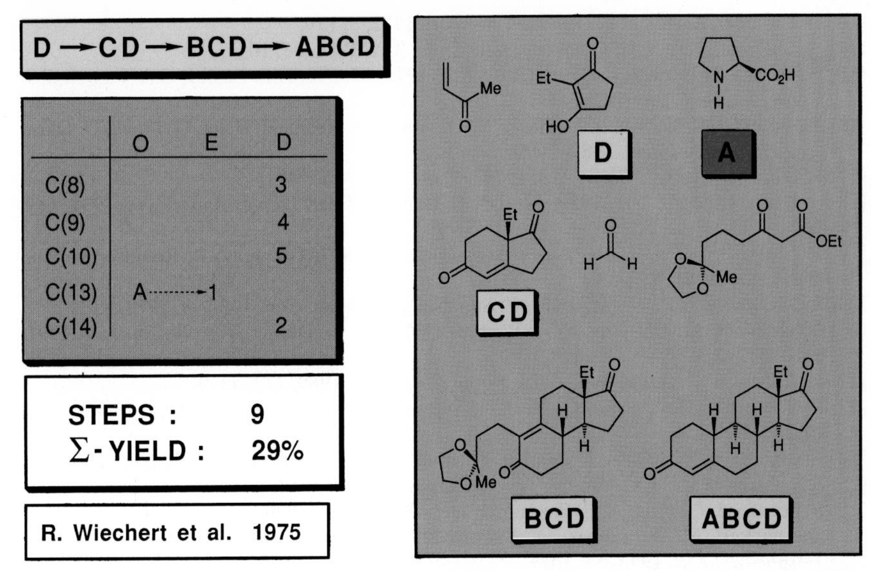

Fig. 30. Evaluation of the steroid synthesis following the constitutional pattern D → CD → BCD → ABCD which gives (-)-norgestrel after nine steps in an overall yield of 29 % (relative to D).

3. Parerga

The historical background of the discoveries of the *Diels/Alder* reaction and the *Woodward/Hoffmann* rules of orbital symmetry conservation has been elegantly elaborated [39]. In a previous essay, the role the *Diels/Alder* reaction has been playing during five decades of steroid synthesis was emphasized [40].

References

[1] O. Diels. K. Alder, *Liebigs Ann. Chem.* **1928**, *460*, 98.
[2] F. von Bruchhausen, H.W. Bersch, *Arch. Pharmaz.* **1928**, *266*, 697.
[3] O. Diels, K. Alder, *Ber. dtsch. chem. Ges.* **1929**, *62*, 554.
[4] O. Diels, *Fortschr. Chem. Org. Naturstoffe* **1939**, *3*, 1.
[5] K. Alder, M. Schumacher, *Fortschr. Chem. Org. Naturstoffe* **1953**, *10*, 1.
[6] O. Diels, S. Olsen, *J. Prakt. Chem.* **1940**, *156*, 285.
[7] J.E. Banfield, D.St.C. Black, S.R. Johns, R.I. Willing, *Aust. J. Chem.* **1982**, *35*, 2247.
[8] K.C. Nicolaou, N.A. Petasis, in *Strategies and Tactics in Organic Synthesis* (Th. Lindberg, Ed.); Vol. 1, P. 155, Academic Press, Orlando **1984**.
[9] W.G. Dauben, C.R. Kessel, K.H. Takemura, *J. Am. Chem. Soc.* **1980**, *102*, 6893: W.G. Dauben, J.M. Gerdes, D.B. Smith, *J. Org. Chem.* **1985**, *50*, 2576.
[10] P.A. Grieco, J.J. Nunes, M.D. Gaul, *J. Am. Chem. Soc.* **1990**, *112*, 4595.
[11] G. Stork, E.E. van Tamelen, L.J. Friedman, A.W. Burgstahler, *J. Am. Chem. Soc.* **1953**, *75*, 384.
[12] G.O. Schenck, K. Ziegler, *Festschrift Prof. Dr. Arthur Stoll*, S. 620, Birkhäuser, Basel **1957**.

[13] D. Lednicer, *Advances in Organic Chemistry* **1972**, *8*, 179.
[14] H. Baier, G. Dürner, G. Quinkert, *Helv. Chim. Acta* **1985**, *68*, 1054.
[15] E.J. Corey, H.E. Ensley, *J. Am. Chem. Soc.* **1975**, *97*, 690; H.E. Ensley, C.A. Parnell, E.J. Corey, *J. Org. Chem.* **1978**, *43*, 1610.
[16] G. Quinkert, U. Schwartz, H. Stark, W.-D. Weber, F. Adam, H. Baier, G. Frank, G. Dürner, *Liebigs Ann. Chem.* **1982**, 1999.
[17] S. Danishefsky, *Acc. Chem. Res.* **1979**, *12*, 66.
[18] E. Dane, *Angew. Chem.* **1939**, *52*, 655; E. Dane, K. Eder, *Liebigs Ann. Chem.* **1939**, *539*, 207.
[19] P. Yates, P. Eaton, *J. Am. Chem. Soc.* **1960**, *82*, 4437; G.I. Fray, R. Robinson, *J. Am. Chem. Soc.* **1961**, *83*, 249.
[20] J. Das, R. Kubela, A. MacAlpine, Z. Stojanac, Z. Valenta, *Can. J. Chem.* **1979**, 3308.
[21] G. Quinkert, M. Del Grosso, A. Bucher, J.W. Bats, G. Dürner, *Tetrahedron Lett.* **1991**, *32*, 3357.
[22] G. Chang, W.C. Guida, W.C. Still, *J. Am. Chem. Soc.* **1989**, *111*, 4379.
[23] S. Scholz, H. Hofmeister, G. Neef, E. Ottow, C. Scheidges, R. Wiechert, *Liebigs Ann. Chem.* **1989**, 151.
[24] I. Minami, K. Takahashi, I. Shimizu, T. Kimura, J. Tsuji, *Tetrahedron* **1986**, *42*, 2971.
[25] S.N. Ananchenko, I.V. Torgov, *Tetrahedron Lett.* **1963**, 1553.
[26] S. Drenkard, J. Ferris, A. Eschenmoser, *Helv. Chim. Acta* **1990**, *73*, 1373.
[27] G. Quinkert, M. Del Grosso, A. Bucher, M. Bauch, W. Döring, J.W. Bats, G. Dürner, *Tetrahedron Lett.* **1992**, *33*, 3617.
[28] E. Dane, J. Schmitt, *Liebigs Ann. Chem.* **1938**, *536*, 196; *ibid.* **1939**, *537*, 246.
[29] G. Singh, *J. Am. Chem. Soc.* **1956**, *78*, 6100.
[30] G. Quinkert, H. Becker, M. Del Grosso, G. Dambacher, J.W. Bats, G. Dürner, *Tetrahedron Lett.*, in press.
[31] W. Döring, *Dissertation to be submitted.*
[32] D. Seebach, A.K. Beck, M. Schiess, L. Widler, A. Wonnacot, *Pure Appl. Chem.* **1983**, *55*, 1807; A.K. Beck, B. Bastani, D.A. Plattner, W. Petter, D. Seebach, H. Braunschweiger, P. Gysi, L. La Vecchia, *Chimia* **1991**, *45*, 238.
[33] M. Bauch, *Dissertation to be submitted.*
[34] W.J. Scott, J.K. Stille, *J. Am. Chem. Soc.* **1986**, *108*, 3033.
[35] C. Rufer, H. Kosmol, E. Schröder, K. Kiesslich, H. Gibian, *Liebigs Ann. Chem.* **1967**, *702*, 141.
[36] G. Sauer, U. Eder, G. Haffer, G. Neef, R. Wiechert, *Angew. Chem.* **1975**, *87*, 413.
[37] (a): U. Eder, G. Sauer, R. Wiechert, *Angew. Chem.* **1971**, *83*, 492;
(b): Z.G. Hajos, D. R. Parrish, *J. Org. Chem.* **1974**, *39*, 1615.
[38] E. Fischer, *Ber. dtsch. chem. Ges.* **1894**, *27*, 3228.
[39] J. A. Berson, *Tetrahedron* **1992**, *48*, 3.
[40] G. Quinkert, Five Decades of Steroid Synthesis, *Vorlesungsreihe Schering*, Heft 19, Berlin **1988**.

Subject Index